On the Production Methods of Pot Still Whisky

Front cover of first notebook

On the Production Methods of Pot Still Whisky
Campbeltown, Scotland
May 1920

Masataka Taketsuru

Translated from the Japanese by
Ruth Anne Herd

Technical editor
Professor Alan G. Wolstenholme

humming earth

Published by
humming earth
an imprint of
Zeticula Ltd
Unit 13
196 Rose Street
Edinburgh
EH2 4AT
Scotland

http://www.hummingearth.com
admin@hummingearth.com

Text Copyright © Ruth A. Herd 2021
All photographs and diagrams courtesy of
Nikka Whisky

Foreword © Alan G. Wolstenholme 2021

Hardback ISBN 978-1-84622-073-9
Paperback ISBN 978-1-84622-074-6

All rights reserved. No part of this publication may be reproduced, stored in a retrieval system, or transmitted in any form or by any means, electronic, mechanical, photocopying, recording or otherwise, without the prior permission of the publishers.

Acknowledgements

Manako Hara was instrumental in getting this project off the ground and provided cheerful and unstinting assistance throughout. Misa Kawamura-Besser checked the translation with meticulous care, providing many valuable insights into the Japanese text. Asahi Beer and Nikka Whisky kindly applied their expertise in the final stages of the project, bringing clarity to those parts that had previously perplexed the translator.

Last, but by no means least, I must acknowledge the enormous help and support I received from Professor Wolstenholme, technical editor of this volume, without which the project could not have been brought to fruition. His enthusiasm was infectious, and he gave generously of his time and expertise to help keep me on the right track.

Warm thanks are owed to all. Any errors and/or omissions are, of course, my own.

Contents

Acknowledgements	*v*
Contents	*vii*
List of Illustrations	*viii*
Translator's Notes	*ix*
Foreword	*xi*

On the Production Methods of Pot Still Whisky	1
Table of contents	3
Preface	5
Raw Materials	9
Production of malt	10
Mashing	22
Cooling	30
Fermentation	32
Distilling	36
1. The distillation process	43
2. Pot ale	45
3. The alcohol meter	47
Glossary	50
The Storage of Whisky	55
The Excise Duty on Scotch Whisky	57
The Price of Whisky	58
Summary	60
Appendix	62

Endnotes	*69*
Further Reading	*71*
Washback Filling Programme	*72*
Measurement Conversion Table	*74*

List of Illustrations

Frontispiece	ii
Table of Contents	2
Map of Scotland hand-drawn by Taketsuru	4
View across distillery roofs	7
Hazelburn Distillery, 1920	7
Map of Campbeltown town centre	8
The malting barn in cross-section	10
Malting floor on level 2 showing the conveyor attached to the ceiling	11
Cross-section of the building, showing the position of the steeping cisterns	12
Storage room on level 4	13
Shovel	14
Two styles of ventilator	16
Two diagrams of the ceramic tiles	16
Thermometer	17
Drying kiln for malted barley	18
Porteus Mill	19
Plant for drying of draff	21
Cross-section of a mash tun	23
Cross-section of device used to mix malted barley and water	24
Table of the Mashing sequence	27
Cooler	30
Cross-section of cooler	30
Bate's Saccharometer	33
The Still and Worm Condenser	38
Enlarged drawing of air valve	39
Mixing Apparatus at the bottom of the still	39
Pipe leading to external worm tub	40
Worm tubs	41
Cross-section of spirit safe	42
Diagram showing the products of wash distillation	45
Sikes' Hydrometer	48
The sample cylinder and the floating weights	49
Technical terms	50
Company housing for workers	67

Translator's Notes

The original manuscript comprised two handwritten notebooks, the covers of which carried the title, "Pot Still Whisky – a Report on Practical Training". The text of these notebooks ran to a total of one hundred and one pages and was interspersed with numerous annotated diagrams, together with a few photographs that Taketsuru himself had taken while in Campbeltown. Taketsuru's neat, fluid handwriting is a thing of beauty to behold and its clarity greatly aided the task of translation. In reproducing Taketsuru's *Glossary*, I have taken pains to employ the orthography originally used. I felt strongly that to replace certain characters with their present-day equivalents would be to rob the text of its historical flavour. Readers will note that almost all the measurements given are native Japanese ones. Conversion tables have been supplied as an appendix for the benefit of those who wish to calculate the Imperial and Metric equivalents. Japanese long vowels have been indicated by a macron, except in the case of well-known place names, such as Osaka and Kobe. The author's name follows the English convention of given name before surname.

There were occasional indications that either a slip of the hand or lapse of memory had led the author astray, for example in his choice of Chinese character, but the number of such instances was vanishingly small. One must bear in mind that Taketsuru had already been operating outside his native linguistic *milieu* for months on end prior to arriving in Campbeltown. Under the circumstances, the odd orthographic aberration is entirely to be expected, and is something for which a translator must make allowance.

The diagrams retain their Japanese keys, which follow the sequence of the *Iroha* syllabary. There are 47 syllables in all, but Taketsuru only uses the first 14 in his diagrams, as shown below. They may be mapped to the Roman alphabet for convenience.

Roman	A	B	C	D	E	F	G	H	I	J	K	L	M	N
Iroha	イ	ロ	ハ	ニ	ホ	ヘ	ト	チ	リ	ヌ	ル	オ	ワ	カ
	i	ro	ha	ni	ho	(h)e	to	chi	ri	nu	ru	o	wa	ka

Foreword

I am delighted to have been asked to provide a foreword to this remarkable work, a translation from Japanese into English of Masataka Taketsuru's notebooks, which run to a total of 101 pages and document the manufacture of Scotch Malt Whisky as practised at Hazelburn Distillery, Campbeltown, in 1920.

The reason that I have been so interested in this translation and happy to help wherever I could is because of a family connection. Peter Margach Innes, my grandfather, was the Distillery Manager at Hazelburn at the time of the study, becoming Taketsuru's mentor during his sojourn there in early 1920. His daughter, my mother, was born there a couple of years later in the manager's accommodation, which was actually within the distillery precincts. Although Peter died long before I was born, it was the family stories about him, and my interest in science, which made me resolve to make a career in the Scotch Whisky Industry myself.

It may well be that many readers will already be familiar with many aspects of the story of the remarkable Japanese gentleman who played such a major role in the foundation of the Japanese Whisky Industry. As, however, it may be some people's first encounter with him, I will share some of the more salient points, without attempting to provide a comprehensive narrative.

Masataka Taketsuru was born in 1894 into a comfortably off *sake* producing family in Takehara, Hiroshima Prefecture. In his teens he attended Osaka Technical High School completing a course focused on alcohol manufacture. He then joined a drinks company called Settsu Shuzo and it was here that a decision was made which would change his life forever. The head of the company, Abe Kihee, was aware of the superior quality of imported Scotch, as compared to locally produced spirit, and decided to send Taketsuru to the other side of the world to Scotland to learn the details and secrets of its production.

He set off from Japan in July 1918 and although the Pacific Theatre was quiet, and Japan was aligned with the Allies, the Great War was still raging in Europe. He travelled by ship to the USA, traversing it by train, and then took another boat, which zig zagged across the Atlantic to deter U boats, eventually bringing him safely to Liverpool. Originally intending to base himself in Edinburgh, he was advised en route that Glasgow might be preferable, and so, in December 1918, he enrolled on a Chemistry course at the University of Glasgow.

During 1919, as well as his lectures, he also took other steps to progress his mission. In April he travelled by train to Elgin where, although unable to agree terms with a notable industry expert (JA Nettleton) for tuition, he was able to spend a few days at Longmorn-Glenlivet Malt Distillery talking to managers, operators and Excise officers.

Back in Glasgow, he took up "digs" (rented accommodation) with the Cowan family in Kirkintilloch and formed a deep relationship with Rita, the eldest daughter of his landlady. This romance culminated in their marriage in January 1920. The story of their lives after the couple's return to Japan formed the basis of *Massan*, a highly popular Japanese breakfast-time TV drama.

He had also managed during 1919 to sign on to an Organic Chemistry course at the Royal Institute (now Strathclyde University) where he was much helped and encouraged by his professor, Forsyth Wilson. The professor seems to have used his established industrial contacts with the Calder distilling family to obtain permission for Taketsuru to spend a few weeks at Bo'Ness Grain Distillery and make a briefer visit to Gartloch Grain Distillery.

He wished however to learn in more detail about the manufacture of Malt whisky and how it was incorporated into the popular blends which predominated in that era. One blend which was of particular interest was White Horse which belonged to Sir Peter Mackie. Whilst Professor Forsyth Wilson was not able to gain him access to the sensitive Blending operations in Glasgow, Taketsuru was allowed to spend an extended period of several months in early 1920 at Mackie's recently acquired Hazelburn distillery in Campbeltown.

Campbeltown had once held a rarified status within the Scotch Whisky Industry as it grew during the 19th Century having a concentrated group of distilleries which earned it the title of "Whiskyopolis". However tastes were changing and blenders were tending to favour the Speyside style for their blends. My grandfather, who had learned his trade thoroughly at Tamdhu and Yoker Distilleries, had been recruited in 1919 by Mackie as manager at Hazelburn in order to make the quality of the Hazelburn whisky more suitable both for "in house" blending and for sale to other firms. Mackie formally announced to his customers that he had "secured the services of a first class Speyside Distiller and Maltman" to improve quality and subsequently congratulated Innes on the improvements he had achieved.

The Scotch Whisky Industry itself was undergoing a period of considerable disruption and consolidation around this time. The Distillers Company Limited, which had started in 1877 as a supply limiting cabal of grain distillers now was swallowing up brand owners and their associated malt distilleries too. Production shortages due to the Great War were exacerbated by steep hikes in domestic taxation arising from both public finance needs and quasi prohibitionist tendencies. The requirement of the 1915 Finance Act that Scotch Whisky must be a minimum of three years old improved quality but increased the working capital requirements. Many distilleries, especially if not owned by a successful brand, sold up or closed down. This was true across Scotland but particularly so in Campbeltown.

Thus Taketsuru might well have considered himself quite fortunate that, after approximately a year of quite sporadic interactions with the industry he wished to understand in its entirety, he found himself, with his new wife, in Campbeltown, and with apparently an indefinite and unlimited access to what was one of the best, if not the best, and certainly the largest, distillery in this epicentre of distilling expertise. Not only that but he was receiving practical tutelage from a highly experienced mentor who was actively engaged in a programme of what might now be termed "continuous improvement" at the site.

Taketsuru had prepared himself for this opportunity well. Although his meeting with Nettleton in Elgin had not proved fruitful, he had acquired a copy of the expert's "magnum opus", *The Manufacture of Spirit as conducted in the Distilleries of the United Kingdom*, and usefully spent many hours in university libraries, as well as Glasgow's Mitchell Library, translating it into Japanese whilst becoming familiarised with the terminology and technology of the processes.

The book you now have the opportunity to read was not intended to be read by the general public. It is a detailed technical manual and blueprint for presenting to his superiors on return to Japan with a view to building an operational distillery.

It is a combination of actual measurements of the production vessels and equipment and explanations of the process steps which would have been clarified by dialogue with an expert, Peter Innes.

It is the book which, together with the undoubted rough drafts used in its preparation, formed the basis for the building of distilleries for three Japanese companies which comprised a significant percentage of that country's now very successful whisky industry.

In addition, it shines a spotlight on the operating practices in Scottish distilleries at a time when little was formally documented.

Distilling arrangements then differed in one particular respect. Nowadays, with a few worthy exceptions, the malting process, from receipt of raw grain through steeping, germination and kilning, is carried out in huge dedicated and specialised maltings. Malted barley is readily available and almost invariably of a very high and consistent quality, whether from commercial maltsters or dedicated central company run facilities. A century ago, again with the odd exception, each distillery produced its own malt, and the maltings capacity was usually the determining factor in spirit output.

Malted barley certainly existed in Japan as beer brewing was being carried out. Taketsuru however appreciated that the particular manner in which Scotch distillers malted their barley impacted product quality and paid particularly close attention to the details of the process.

Taketsuru, in his report, shows himself to be more than just an accurate observer. His scientific knowledge, curiosity and creativity are also evident. He also shows himself interested in the characteristics of the labour force, comparing them with their counterparts in his homeland. He also clearly shows the central role His Majesty's Customs and Excise (now Her Majesty's Revenue and Customs) held in controlling the key activities in order to protect the enormous amounts of duty which would become liable. Again with regard to finance, he calculates what would now be called a "cost of make" for the spirit he intended to produce.

After leaving Campbeltown in May 1920, the Taketsurus returned to Japan. The good copy of this notebook was presented by Taketsuru to his immediate superior, Iwai Kiichiro, at Settsu Shuzo. However the mood had changed there regarding producing Scotch type whisky and, shortly afterwards, Taketsuru left the company. Rita and he both took up teaching roles at St. Andrews College in Osaka.

No more might have been heard of this episode, nor even possibly Japanese whisky, were another remarkable Japanese businessman not to enter into this story. Shinjiro Torii was an entrepreneur who had built up a successful business selling Akadama wine, a sweet offering which suited Japanese tastes. His company, Kotobukiya, became in time the international behemoth, Beam Suntory, now such a major player in global spirits sales.

Torii had also identified the opportunity that would arise from producing whisky locally. Initially he thought that he would have to lure an expert over from Scotland to assist him before learning that help was closer to hand.

A deal was agreed whereby Taketsuru signed a ten year contract to design, build, commission and manage Japan's first whisky distillery, Yamazaki, which opened in 1924. The equipment, including the distillation apparatus was manufactured in Japan to Taketsuru's specifications. Using the information learned in Scotland he produced whisky spirit and in 1925 returned to Scotland with a sample of Yamazaki "new make" to discuss its quality with his mentor, Peter Innes.

A further visit to Scotland and meeting took place in 1931 when Taketsuru was entrusted with escorting

Torii's heir, Kichitaro Torii, to Scotland to expand his knowledge. By this time the end of the ten-year contract was in sight and Taketsuru's thoughts were turning to starting up on his own.

His knowledge, notes and diagrams again came to the fore and formed the basis on which he built a distillery for himself at Yoichi in Hokkaido, a location chosen for its similarity to Scotland in general and, some say, Campbeltown in particular. From this start the other major Japanese whisky company, Nikka, grew.

The "good copy" of the original notebooks stayed in the possession of Settsu Shuzo. Some years later, it was used as a guide when Hombo Shuzo was setting up its whisky production. Subsequent to this it was donated to Taketsuru's son and now has pride of place in the Yoichi Distillery museum. A reprint in Japanese was at some point carried out for the benefit of Japanese whisky enthusiasts. By chance, a Japanese friend of the translator, knowing her to be a native Campbeltonian, made her a gift of the facsimile notebooks and that is how this translation has come about.

After Taketsuru's stay at Hazelburn, Peter Innes worked with Stuart Hastie, "the first distillery chemist" on a research programme about the science of whisky manufacture, which was subsequently published in the highly respected *Journal of the Institute of Brewing*. In 1923 Peter Mackie acquired a controlling share in Cragganmore Distillery in Speyside and Peter Innes was transferred there to bring it too to a higher standard. Hazelburn Distillery did not survive the contraction of the industry very much longer. By 1925 the distillery was shut for spirit production though the bonded cask warehouses were operated by Distillers Company Ltd, who purchased White Horse, for another half century. All that remains of Hazelburn distillery is the accommodation and office block which is now a business centre.

The name and tradition of Hazelburn whisky thankfully lives on as a brand of the admirable J&A Mitchell & Co and is made by very traditional methods at their Springbank distillery, only a matter of yards from the original Hazelburn Distillery location.

Taketsuru's text shows him to be thoughtful, detailed and insightful as he goes about his task of creating a manual on how to set up a malt whisky distillery from scratch. Measurements, weights, volumes, times and temperatures are all diligently recorded using the units his colleagues would understand. He doesn't only observe the normal running of the distillery but also set up and monitored a "mass balance" type experiment from a standing start situation. The success of Taketsuru's mission would depend, as well as an observation of activities and opinions of those he encountered, on the accuracy of the data regarding the buildings, equipment, raw materials, ratios, temperatures and yields he objectively recorded.

Japan had an ancient and effective system of weights and measures. The "opening up" of Japan in the 19th Century had led to an awareness, and partial adoption, of the Metric system.

Nothing would have quite prepared the young Masataka Taketsuru for the UK's Imperial System which was in use in Scotland and across the eponymous Empire.

Bushels, Fahrenheit and yards would be challenge enough but worse to comprehend were those related to measuring density and alcoholic strength. In particular the Sikes hydrometer was an instrument which he had to come to terms with, as it provided results in an idiosyncratic "proof gallon" format. (I had to use this system for the first decade (1970s) of my career in distilling, as it was only phased out as late as 1980, finally giving way to the logical and consistent metric (OIML) system).

Similarly, the use of Bates *Saccharometer* to measure wort gravities, rather than the Balling system he had been taught, was a major challenge for him.

Many of the Imperial system's idiosyncrasies also feature in the measurement system which the USA used and still use, although there are a few traps for the unwary, such as the smaller American gallon!

Taketsuru recorded his findings in the Japanese units, and that is what the translation retains. However, for better understanding, the reader will wish to appreciate the content in familiar units, and having made that journey, I wish to share that in the relevant appendix, mostly in tabular form, and suitably but approximately rounded.

I came across an occasional calculation where I struggled to make sense of Taketsuru's workings, but they are few and far between and it would be carping to shine a spotlight on any in particular.

However, underneath his writings which were recorded in units suited to his colleagues' understanding, can be discerned a past ghostly world of a vanished period of the history of the Scotch Whisky Industry, sketched out in proof gallons, square yards and bushel weights, which does not deserve to be completely forgotten.

For example, back calculation shows that his mass balance was conducted from a grainbill of exactly 1000 bushels!

(Out of interest I converted his *monme* figures to lbs and then metric tonnes etc. and from his reported results calculated that the distillery was achieving approximately 313 lpa/tonne from its Canadian barley malt. This is three quarters of the approximately 420 lpa/tonne considered achievable today, showing the much greater yield potential of modern barley strains).

He projects his thoughts forward to the challenges of setting up a distillery where little of the necessary infrastructure exists and also thinks creatively about how capital equipment can be improvised and costs kept to a minimum.

As well as the equipment and raw materials needed he also considers the manning which will be required, and although his terminology was not quite as would be used today, he shows himself to have been a progressive thinker of his time advocating improvements in working arrangements for prospective employees.

I don't always agree with every one of his technical proposals; some of his production schedules look optimistic and a couple of his calculations, in my opinion, contain anomalies. In working with Ruth to produce this translation we have had many debates about how we might phrase something today but have tried not to stretch anything very far what he actually said. However what shines through these pages is an enthusiastic and intelligent young man who is determined to take the opportunity fate has offered to absorb information regarding the industry and product he is passionate about. The knowledge he gleaned at Hazelburn Distillery and elsewhere in Scotland set him up for the career he pursued successfully for the remainder of his life and without which the Japanese Whisky Industry would not be the international colossus it is now, a century on.

Professor Alan G Wolstenholme
International Centre for Brewing and Distilling
Heriot Watt University, Edinburgh

On the Production Methods of Pot Still Whisky

目次 (一)

	頁
序	一
原料	五
麦芽製造	七
一、麦芽製造室の構造	七
二、大麦の浸水	一三
三、発芽	一五
四、乾燥	二〇
五、麦芽粉砕	二三
糖化	三二
一、糖化日割	三七
二、糖化用機械器具	四〇
三、糖化作用	四四
四、糖化残渣	五五
冷却	六〇
醱酵	六四
蒸溜	六七
一、蒸溜操作	六七
二、蒸溜残渣	六七
三、アルコールメーター	七五
四、術語	七八
ウヰスキー貯蔵に関する事項	
ウヰスキー税額	

Table of contents

Preface	5
Raw Materials	9
Production of malt	10
1. The structure of the malting barn	10
2. The steeping of the barley	13
3. Germination	14
4. Drying	15
5. The milling of the barley	19
Mashing	22
1. The mashing schedule	22
2. The device used for producing wort and its apparatus	22
3. The effect of mashing	24
4. The draff	26
Cooling	30
Fermentation	32
Distilling	36
1. The distillation process	43
2. Pot ale	45
3. The alcohol meter	47
Glossary	50
Storage of whisky	55
Tax/Excise	57
Pricing	58
Cooperage	60
Summary	60
Appendix	62
Labour issues	62
Pay and conditions	65
Methods of sale	66

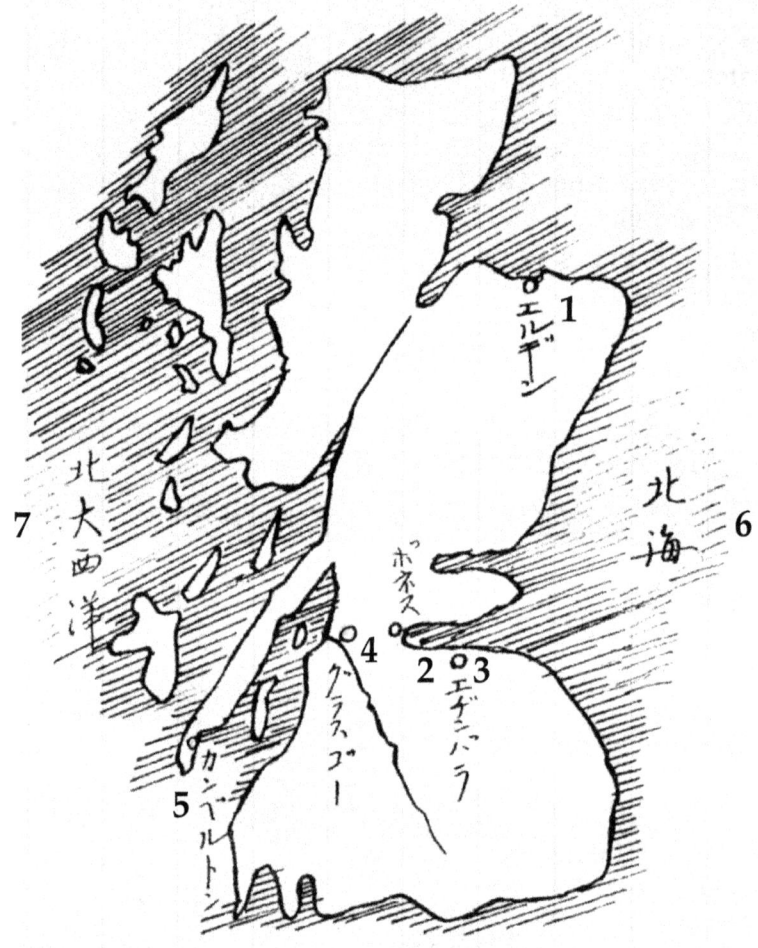

Map hand-drawn by Taketsuru showing the positions of
Elgin, Bo'ness, Edinburgh, Glasgow and Campbeltown, and tracing the course of the River Clyde from its source.

1. Elgin
2. Bo'ness
3. Edinburgh
4. Glasgow
5. Campbeltown
6. North Sea
7. North Atlantic Ocean

Preface

After undergoing a period of practical training in the northern city of Elgin in the April of last year, I produced a report summarising the production methods of pot still whisky. That report was somewhat incomplete. It is therefore my intention herein to rectify those shortcomings. At the same time, I aim to go a step further and set down an account of the production processes that may serve as a manual on the basis of which to implement this style of whisky production in Japan.

Within Scotland, each region produces a whisky with its own distinct bouquet. Elgin is representative of the Highland whisky of the north, while Campbeltown stands by itself and is recognised as having its own distinctly "Campbeltown" character. This report deals with the methods by which the latter, i.e. *Campbeltown* whisky, is produced. Firstly, I look at the geographic features of Scotland and identify the various similarities that the two types of whisky share. I feel that there is something to be gained by making such a comparison.

Elgin lies facing the Moray Firth, an inlet of the North Sea. One might think that winters there would be as cold as in northern Japan, but the region benefits from the influence of the warm ocean current that flows all year round. Therefore, while the Central Plain may still be covered in snow, in Elgin the flowers are blooming and the larks chirping. It is a marvellous sight.

To the south of Elgin lie the Grampian Highlands[1] and what are referred to in the Scots dialect as 'burns', the small streams that provide the water needed to produce whisky. Campbeltown lies to the south on the Firth of Clyde, with the mountains of the Argyll Range at its back. Cold winds blow in from the North Atlantic on the west side, but the hills keep the climate in temperate balance. The famous whisky made in Campbeltown is produced from a stream of water drawn from Crosshill Loch, which lies high up on one of the mountains of the Argyll Range.[2]

Generally speaking, distilleries engaged in the production of pot still whisky are located facing the sea. When I asked a local expert, I was told that, in the case of continuous distillation, it would be very difficult to obtain a good result unless the factory were situated at a distance from the sea, as the two methods require quite different topographical situations in order to be successful. It is hard for me to understand this reasoning but, sure enough, when I examine the position of the factories engaged in continuous distilling in the UK, I find that they are indeed almost all located at quite a distance from the sea. In Campbeltown, which faces the sea and where I am now doing my training, there is not one factory engaged in continuous distilling – they all use the pot still method.

Campbeltown lies on the west [sic] side of the Kintyre Peninsula, roughly five hours' journey by steamship from Glasgow down the River Clyde. Apart from being one of the country's foremost centres of whisky distilling and fishing, all year round, and especially in summer when visitors are at their most numerous, the town boasts the most wonderful scenery and a good harbour. In the past there were more than twenty distilleries but seven became amalgamated so that now only fourteen remain, lined up across the western side of the town, in competition with one another for survival both day and night.

I received permission to train at the largest of the town's distilleries – Hazelburn. Mr Innes, the chief technician, was extremely kind and assisted me in all matters. This is the Mackie distillery that produces the famous "White Horse". There are four distilleries in Scotland that produce "White Horse". Three use the 'pot still' method: Craigellachie in Elgin, Islay in the south [sic], and the one in Campbeltown. Then in Glasgow there is a continuous distilling factory that uses grain other than barley to produce odourless spirit that is then blended with the product of various pot stills, after which it is sent to market.

At the end of last year, I repeatedly asked to be given permission to do some training at the Mackie Company in Glasgow but even today have still not received such permission. Not wishing to waste the precious opportunity of having come to Scotland, and keen to do some training at a prestigious distillery, I ran around all over the place and eventually an acquaintance introduced me to contacts in Campbeltown, and that is how I came to be here. Now, as the time for my return to Japan draws close and I have at last fulfilled my purpose, it gives me enormous pleasure to be able to offer this report on the distilling methods of "White Horse".

Opposite, top: View of Campbeltown, looking east over distillery roofs towards Davaar Island.
Lower: Hazelburn Distillery

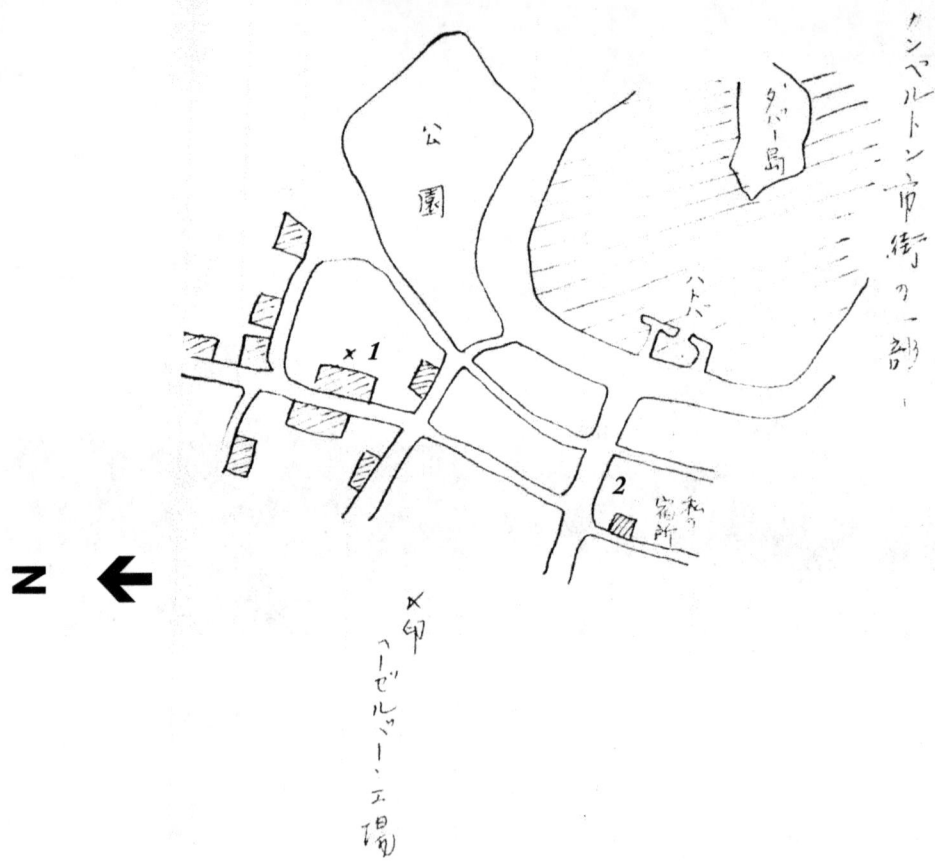

Map of Campbeltown town centre showing (1) the location of Hazelburn Distillery, and (2) Taketsuru's lodgings in Kirk Street.

Raw Materials

Even in the continuous distilling process, the selection of raw materials is an extremely serious business, and it goes without saying that this is also of crucial importance in the case of pot still whisky. When I first came to Scotland, I was under the impression that it was the barley produced in Scotland that gave the whisky its character. Now I realise that Scotland is not a country distinguished by its agriculture. It might be, were it not for the fact that the population is sparse and there are few farmers. Moreover, a large workforce is required to work in their coal mines, which may be said to be the best in the world. That is why, in the hills that stretch out as far as the eye can see, there are mostly livestock farms that require little in the way of manpower. The barley that is produced in Scotland has extremely large grains and is of good quality. Canadian barley simply cannot compare. However, Scotland does not produce enough barley to satisfy the needs of its distilleries and so has to import barley from the neighbouring island of Ireland. Irish barley resembles Scots barley in having large grain heads and it is of good quality, too. Unfortunately, even with imported Irish barley there remains a shortfall, so that more than half the barley used in Scotch whisky production has to be imported from Canada.

As for Canadian barley, on average they use 300 *monme* for every *shō*. After sprouting and drying, 210 *monme* are obtained. If Scottish barley is used[3], then an average of 360 *monme* give 258 *monme* after sprouting and drying. From this, the difference in quality between the two types of grain can clearly be seen.

The quality of Canadian barley somewhat resembles that of Japanese barley. If Canadian barley can be used in whisky making, then there is no reason that Japanese barley could not be. This is a welcome discovery, as we had been under the impression that Japanese barley would be of no use. We can now proceed to give it a try with every hope that we will succeed.

Production of malt

1. The structure of the malting barn

Most of the distilleries in Scotland are broadly similar, with little variation in terms of their equipment and modes of operation. I shall for the most part be describing the equipment used at Hazelburn Distillery, but it may be taken to be representative of the rest. The barley brought into the distillery is moved by elevator to the fourth level, from where it is dispensed as needed. Thus 'vertical' operation commences. I have shown this in an illustration.

The barley that has been transported from point （イ）is sent by elevator （ロ）to the fourth level and is then sent along the screw conveyor （ハ）until it arrives at its destination, where it is to be stored. The conveyor is cylindrical in shape and made of iron, and contains within it a single axle attached to a screw. The screw conveyor transports the barley as the axle turns. There are several chutes attached to the conveyor and it is constructed in such a way as also to allow for the transport of the barley to the floor of the third level.

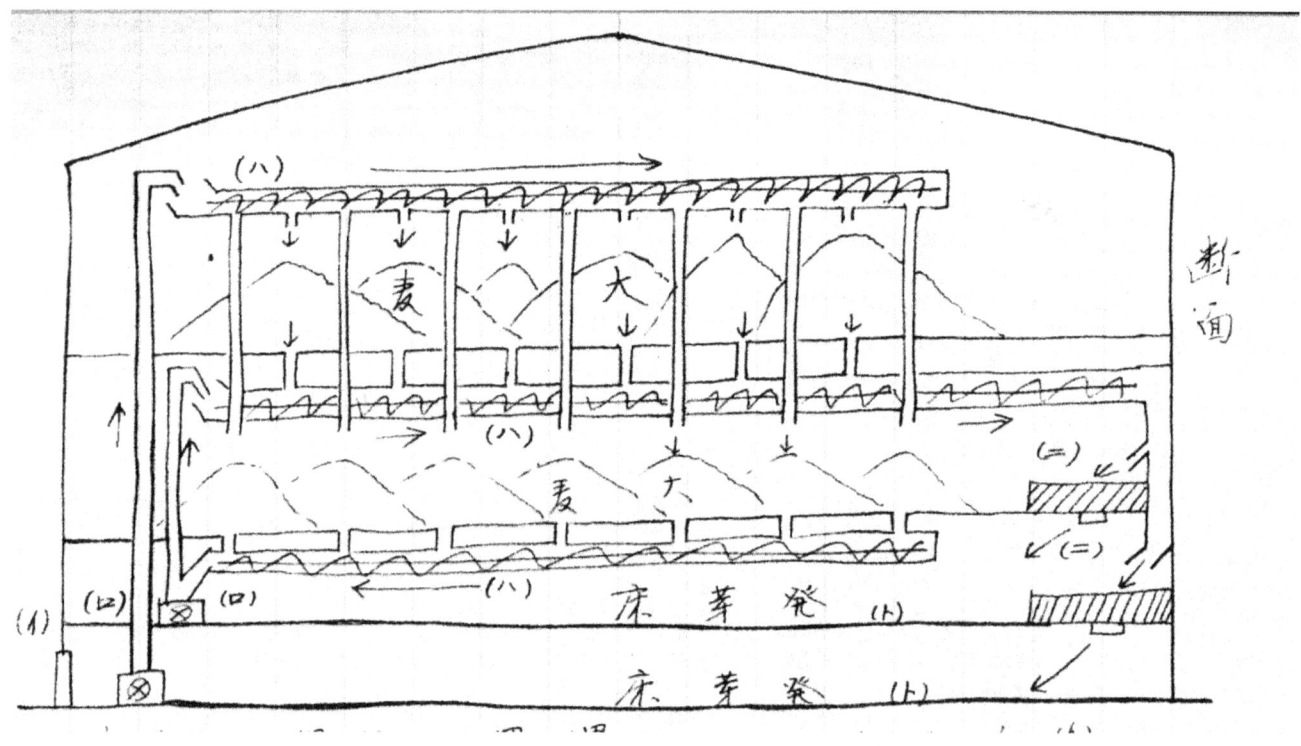

The malting barn in cross-section.

Malting floor on level 2 showing the conveyor attached to the ceiling.

(二) refers to the two steeping cisterns, which are installed on the second and third levels. Since the construction is large in scale, it is necessary to have a conveyor in order to transport the barley to the tanks. The conveyors on levels 2 and 3 are there for the sole purpose of conveying the barley to the steeping cisterns. Holes have been drilled here and there in the floors of levels 3 and 4, connected to the conveyor by pipes (chutes). The volume of barley in the pipes can be adjusted by means of a stopper, which can be opened and closed as required. The conveyor on Level 2 takes the barley from Level 3 to the conveyor and thence to the cistern by means of the elevator (口).

(ト) shows the way in which the barley from the steeping cistern on level 3 is then spread out over the malting floor of level 2, while that in the steeping cistern on level 2 is spread over the ground level malting floor. This structure stores barley on levels 3 and 4, while level 2 and the ground level form the malting floors. As a result of being able to move the barley entirely through mechanisation in this way, the workforce can be small in number and the operation can continue day after day. The building is 4 *ken* x 15 *ken*, and the malting floor is concrete, while everything else is made of wood. On the walls are a number of horizontal pivoted windows by means of which the temperature inside may be adjusted. The walls are extremely thick and minimize the effect of the external air temperature. However, there is nothing in the way of cooling or heating pipes set up inside to adjust the temperature. Production of barley malt differs from that of *kōji* (rice malt) in that, in the winter, production is possible even without the installation of heating pipes indoors, and in summer production ceases as both barley and the water needed for distilling are in short supply — so there is no need for cooling pipes in summer either. Small paths have been created in the centre of the malting room and as the workers walk to and fro they do their best not to tread accidentally on the germinating barley. The workers wear rubber shoes or shoes made of canvas.

The dimensions of the concrete barley steeping cistern are 1 *ken* by 4 *ken* and 4 *shaku* high (about 100 *koku* capacity). At its base, there is a pipe into which numerous small holes have been drilled. Through these, compressed air and water can be introduced as the barley is being steeped. The cistern has in the centre of its base a trough with an iron plate into which small holes have been drilled. These allow the water to be released, whilst preventing any barley from escaping.

My thoughts on construction at the main plant in Japan with a view to the production of malt

No matter what the enterprise, and even when building a distillery, starting from scratch on a large scale with an abundance of resources has the virtue of speed of operation, and of convenience. However, when there is little expectation of success it can lead to heavy losses. Based on what I have hitherto seen of the processes by which distilleries are constructed, I can report that almost all have expanded gradually, keeping pace with the development of the enterprise. For that reason, I believe that, while such distilleries might look unsightly, in order to ensure the solid development of the enterprise, there is no alternative but to follow this path. That is why I would urge that Head Office start off on a small scale.

I believe that it would be best to proceed with construction as follows. Firstly, build a three-storey wooden structure and spread concrete at ground level to form a malting floor. On the second and third levels have the storage area for the barley. I believe that the most convenient location for the steeping cistern would be on Level 2. Provision of an elevator is essential in order that barley can be transported to level 3 but, if the distillery is small in scale, a conveyor will not be required.

If the top of the cistern is set level with the floor of level 2, it will be a simple matter to transport the barley on level 2 to the cistern. At the same time, two holes should be drilled in the base of the cistern, so that when the cover over the holes is opened the barley will automatically transfer itself to the malting floor. If it is not possible for reasons of cost to install a means of introducing compressed air to the tank, then the water should be made to lie as shallow as possible with as large a surface area as possible. Pains must be taken to ensure that the barley can absorb air.

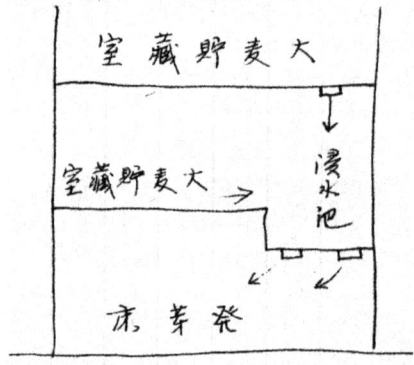

Cross-section of the building, showing the position of the steeping cisterns.

The dimensions of the malting floor should be approximately 4 *ken* x 15 *ken* and in that case one may expect to produce 67 *koku* of malted barley per week. However, it is difficult for me to set down a record of construction because one cannot confine oneself to constructing a malting room. One also has to build a drying room, milling room, storeroom and so on. The foregoing has been my report on construction. Next, I should like to give an account of the production of malt.

2. The steeping of the barley

The barley that has come down from the store into the cistern is not immediately flooded with water. Approximately eight hours later, water is delivered through the pipes at the base of the tank. This is to make sure that, during that time, the barley has had sufficient contact with the air. As the barley will swell by 25% when it is soaked, it is necessary that a surplus volume of 30% be made allowance for when constructing the receptacle. For that reason, the cistern should be 24 *shaku* in length, 6 *shaku* wide and 4 *shaku* in height. The volume of water should be 48 *koku*.

The part of any whisky distillery that is largest in area is the malting barn. Hazelburn has six cisterns. For each steeping cistern an area of malting floor equal to 60 *tsubo* is made available. For that reason there are three buildings standing side by side, each with the first and ground floors given over to use as malting floors, so that the entire operation extends to 360 *tsubo* of malting floors, and in a two-week period 800 *koku* of malted barley can be produced.

Storage room on level 4.
X = chutes used to convey the barley from the conveyor to level 3.

An hour after the steeping of the barley has commenced, the dirt and broken grains of barley[4] that have risen to the surface are scooped out, and the temperature of the water is then kept between 50 and 60 degrees Fahrenheit (10-16 degrees Centigrade) while the barley steeps for 48 hours. During that time, the water is changed three times, and from time to time compressed air is introduced via the pipes, thus preventing the barley from suffocating. Steeping is concluded once the grain is soft enough to be squashed between the index finger and thumb. These processes are familiar to all engaged in distilling, so there is no need for me to repeat them here. The length of time during which the barley is steeped and the number of changes to the water all require adjusting in response to differences in the quality of the barley and the climate. As far as these aspects are concerned, therefore, it is not necessary to replicate the process as it obtains in Scotland.

3. Germination

Firstly, the barley should be spread out across the malting floor to a depth of 1 *shaku* 5 *sun* and, as far as is possible, the temperature kept at an even level. After about 13 hours, the germ will have begun to appear[5] and at that stage the barley should be turned over and mixed using a shovel. The temperature at this time will be between 13 and 16 degrees Celsius. In the UK they use only Fahrenheit thermometers, but I have converted the figures into Celsius for the purposes of this report.

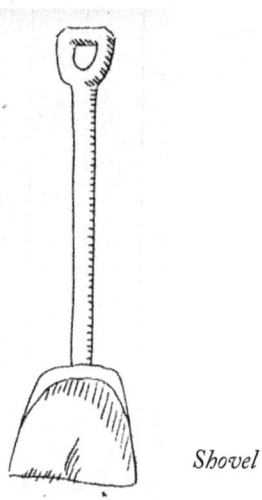

Shovel

The shovel is made entirely of wood so as not to damage the barley. The turning is not done in a random fashion, but rather moves outward from the place at which the barley is dropped down from the cistern. In this way, after several turnings, a space will have opened up below the cistern, into which the next batch of barley will fall. In order to make adjustment for a rise in temperature, after the second turning the depth is maintained at 1 *shaku*, that is to say, slightly shallower than was the case for the first turning. The temperature is thus kept at roughly 20 degrees, and germination may proceed.

In this way, the leafbud[6] inside the barley reaches 3/4 of the size of the full-length germ and, once the rootlet has grown to roughly one and a half times the length of the germ, the process is complete. The process takes between 12 and 14 days, and on the final day the barley will be lying extremely shallow, at around 6 *sun*.

The malting of the barley is the most important of all the manufacturing processes in whisky distilling, and it is no exaggeration to say that the quality of the final product is determined in large part by the quality of the malt. Therefore each distillery has its own distinct method of dealing with the malting process. Take the brewing of *sake* as a similar case in point. There, too, the most important thing is the production of *kōji* (rice malt). Each firm carries on its own tradition in that regard. For example, it would be extremely difficult for provincial Japanese *sake* brewers to obtain good results were they to copy the methods used at Nada Gogō.[7] In the same way, it would be extremely risky for a Japanese distiller to copy methods originating in Scotland, several thousand miles distant.

For that reason, even though this is the most important of all the processes of whisky distilling, I am afraid my report has to remain rather basic and I will be unable to provide anything more in the way of detail. However, I would like to try improving the quality of malted barley produced in Japan through the practical application of these techniques over time.

4. Drying

The drying method is consistent throughout Scotland – that is to say, a drying kiln is used. The furnace must be connected to the malting barn, and it is constructed in such a way as to make it convenient to move the malted barley there. The fact that the kiln is connected to the germination room and barley storage area means that, at first glance, it appears to be a fire hazard. However, since the kiln is, in fact, surrounded by iron plates or heat-resistant tiles, and since the fire does not burn ferociously – and on top of that there being more than a sufficient distance between the grain to be dried and the fire - there is no danger associated with this process.

The area of the drying floor is 5 *ken* square, and the walls surrounding it are made of clay and extremely thick. There are two or three glazed windows, and on the outside at the top there is a means of ventilation. In Elgin in the north, the ventilator is as appears in the illustration on the next page .

The roof is formed in the shape of a pagoda and is between 6 and 8 *shaku* high. It lets the humid air from the inside out naturally, and allows the fire to burn fully. In Campbeltown, however, a revolving device is installed on top of the roof, which turns according to the direction of the wind. It is extremely short in length and placed only about 1 *shaku* above the roof. If the ventilator allows the air on the inside to be expelled with too much force, the fire in the furnace may become fierce, and sparks may fly onto the barley and set it alight. Therefore, owing to this risk of fire, the use of overly efficient ventilators must be avoided.

The drying floor is positioned at a distance of roughly 20 *shaku* from the furnace and has several iron bars affixed to it, and above them, as shown by the illustrations, tiles measuring 2 *shaku* square and with a thickness of 2 *sun* are laid out and fixed in place with cement.

In the space between the furnace and the drying floor there is an iron plate measuring 2 *ken* square. Heat rising from the furnace is blocked by the iron plate, and in this way any sparks are extinguished, preventing them from

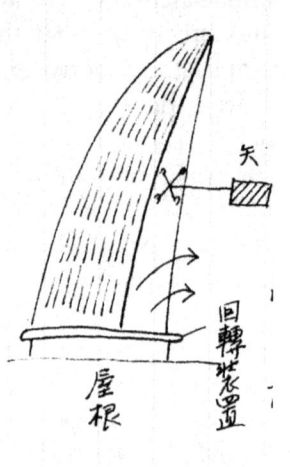

Two ventilators.

Campbeltown style of ventilator

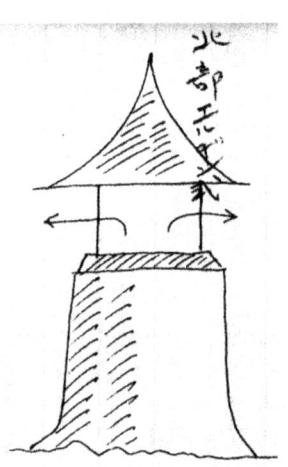

Elgin style with the pagoda roof

rough handling as the men carry out their work, the workers tell me that breakages are rare and that the tiles seldom have to be replaced. If there is a fire, normally this is because of imperfections in the construction of the kiln. If the floor is not strong enough, therefore, there is a concern that the whole floor might suddenly collapse during the drying process and cause a fire.

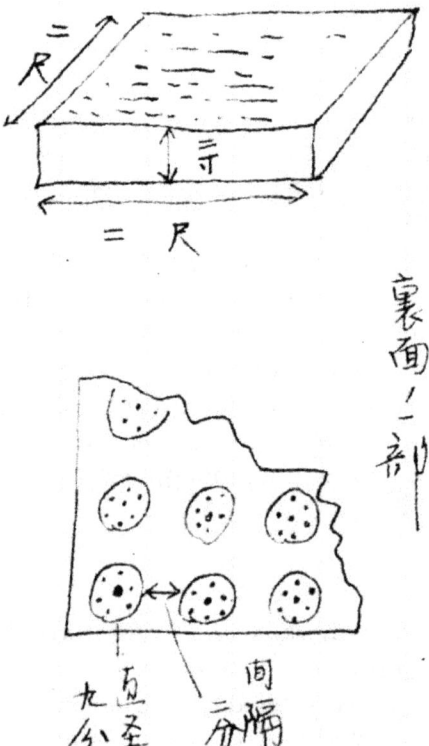

Two diagrams of the ceramic tiles. The bottom one is a cross-section of the reverse side.

reaching the floor. The plate also functions as a means of supplying heat to all parts of the floor. The holes on the front-facing side of the tiles need to be smaller than the grain heads, while the holes on the reverse side have a diameter of 9 *bu*. They are spaced at a distance of 2 *bu* apart. The tiles are extremely strong and even if they come in for

Peat or coke is used in the furnace and so there is no need for a chimney. At Hazelburn, instead of coke they use a mixture of Welsh smokeless coal and peat. Scotland is famous for producing coal but it cannot produce coal with the special qualities of Welsh coal. Not only is Welsh coal smokeless, it is said to imbue the malt with its characteristic whisky aroma. As I deepen my research into the details of raw materials and manufacturing processes, my thinking has come increasingly to the conclusion that it will be extremely difficult to replicate Scotch whisky in Japan, where we are unable to produce peat or Welsh coal.

Next in the drying process is for the barley that has been moved to the drying floor from the malting floor to be spread out to a depth of roughly between 9 *sun* and 1 *shaku*. If the drying causes the temperature of the barley to rise suddenly, then what is known as "glass malt" is produced. Therefore pains must be taken to ensure that the temperature starts off low at about 30 degrees and then, once the humidity falls slightly, to increase the temperature to 60 degrees C and maintain that level. At a temperature of 63 degrees, drying is concluded. While this is taking place, the barley should be mixed, and mixing should take place 3-4 times over a period of roughly twenty-four hours. The thermometer used for the drying process is shown in the illustration. It is placed above the tiles on the floor at a fixed height, and covered with wire mesh or copper plate in order to protect against breakages. After the drying process is complete, good barley malt can be easily crushed between the teeth. It will taste sweet and give off a pleasant aroma.

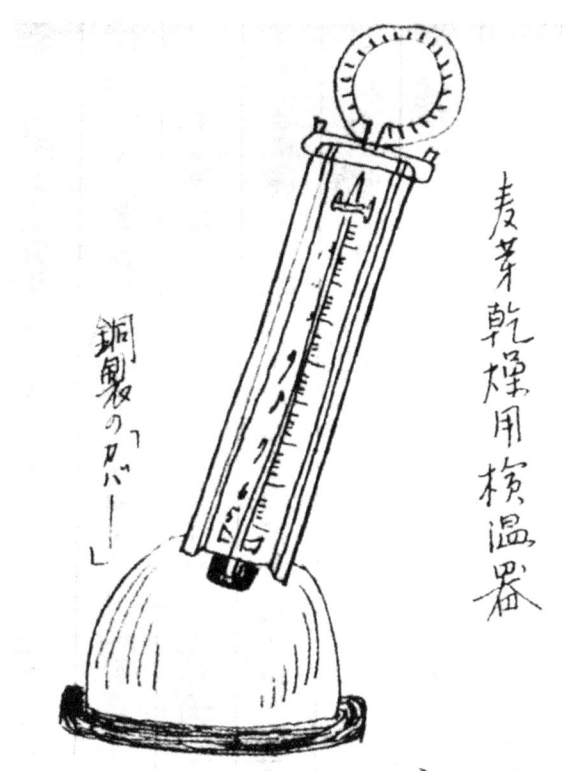

Thermometer (the copper cover at the bottom of the device is pointed out).

During this process, the kiln has reached a temperature of 63 degrees, and so the fuel should not be raked but rather left to cool naturally for 5-6 hours. Only then is the barley to be placed in sacks with a capacity of 13 *kan* and taken to the storeroom. The malted barley will not be taken for mashing immediately after it has been dried, but put into a mill where it is milled and, at the same time, the culms are removed.

My view on the practical application of a drying kiln at Head Office in Japan

Throughout Scotland, only dried malted barley is used for mashing, never green. The reason a drying kiln is used is firstly that the British Isles get a great deal of rain, and in winter sunlight is very weak so that an artificially-created drying process is required.[8] Secondly, local peat is used to impart a unique aroma. Thirdly, through drying, the culms, which have a bitter taste, are removed and the grain is able to withstand a lengthy period in storage.

The fact that peat cannot be produced in Japan would seem to constitute a death blow to any hopes of producing Scotch whisky, with its distinctive aroma, there. Nevertheless, it simply will not do to do away with the drying kiln and dry under the heat of the sun instead. Even if peat cannot be used, it ought to be possible to use a drying kiln in such a way as to produce a flavour similar to Scotch by bringing out the aromatic characteristics naturally inherent in the grain.

Since it is necessary to keep the temperature low at first in order for there to be a strong mash, in a climate such as that of Japan the malted barley should first be dried in the sun and only then put in the kiln. This would save on fuel, and it is something that I would urge us to experiment with.

The only suitable fuel we have in Japan is coke. The structure of the kiln has already been described earlier but here I have provided a simplified illustration. There is a four-cornered square brazier of 1.5 *ken* on which the coke is burned. There is no need for iron plates – if the walls are made of the thickest and most fire-resistant tiles, then the kiln will be safe. This is the kiln I saw in use when I visited Bo'ness last year.

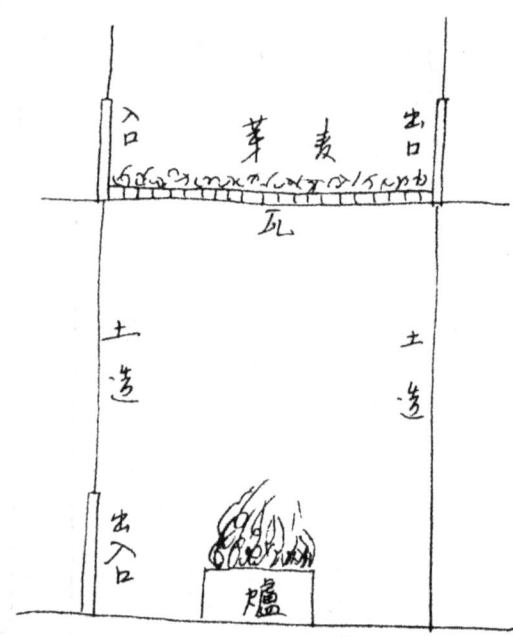

Drying kiln for malted barley.

The kiln is by no means difficult to build but, before construction takes place, I would recommend visiting one of our Japanese breweries to observe how they go about drying their grain when they make beer, to see if there is anything to be learned from them.

Naturally, when it comes to drying grain, the process that applies to distilling whisky will differ from that used in brewing beer, but the difference will lie in the length of time during which drying is carried out; generally speaking, we should find that the process is largely similar. In my opinion, when it comes to building a distillery, rather than apply the fruits of new experiences, we need to take note of the long-established practices in Japan that take account of local characteristics. Then we should improve those aspects that are relevant to whisky production. In that way we will achieve optimum readiness and results. However, that is by no means to imply that we are obliged to follow their methods, merely that we should make use of their knowledge insofar as it is useful to us.

5. Milling of the malted barley

The milling process is not important enough to demand a special section of its own, but I wish to present my research into the mill itself. This style of mill is not tailor-made but comes as standard when ordered from the manufacturer. At Hazelburn, they have installed the most up-to-date model - which has been patented – the Porteus Mill.

As can been seen from the illustration on the right the casing consists of a box made of iron plates into which a window has been cut so that the inner workings are visible. The dried barley drops down through the funnel (イ) and then through the action of the fluted rolls (ロ), the first milling takes place. The apparatus at (ハ) is encircled by countless small apertures and, after passing through the fluted rolls (ロ), the milled powder comes out through these perforations. (ニ) indicates the position of the brushes. As they turn in the directions indicated by the arrows, they compress the powder and squeeze it out. That which does not pass through the apertures

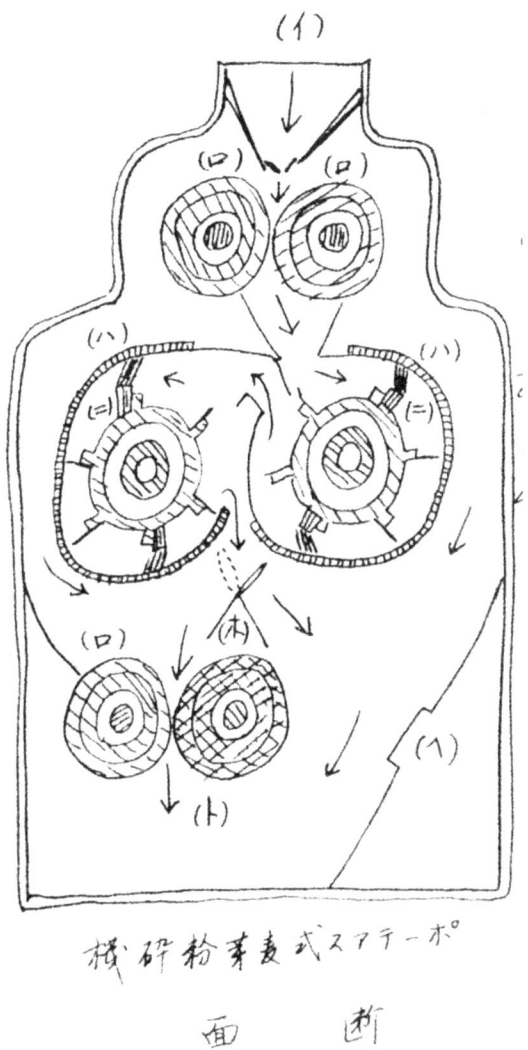

Porteus Mill

ends up on the roller on the left. The openings around the left-hand roller are slightly larger than those on the right so there is a possibility that items other than husks may pass through them. The husks are removed by means of the brushes, but sometimes the workers take some from the sample opening (ヘ). If large grains that require milling are found among them, the levers (ホ) are turned to the right, as shown in the diagram, and then the grains are passed through the fluted rolls (ロ) for a second time. The process having been repeated, the grain is now thoroughly milled. At point (ト), the material that has collected from various places is mixed thoroughly and, when it emerges from the machine, the milled malted barley is now in a condition to be used for mashing. The machine is 9 *shaku* in length, 6 *shaku* high and 4 *shaku* wide, and is driven by means of an external belt.

This machine is expensive owing to its high degree of complexity. Therefore to buy such a machine from the UK in order to experiment in whisky distilling on a small scale would be not be justifiable. What is required is the milling of the grain, and the manner in which this is carried out does not affect the quality of the end product. For that reason, I believe that Head Office should first pass the dried grain through a very fine metal sieve placed on a slant. When the sieve is shaken to the left and to the right, the culms will pass through the mesh and thus be removed, so that only clean barley is sent into mill. Moreover, if, in addition, fans could be placed at the side to blow anything unwanted that has not passed through the mesh away to the opposite side, the process would be complete. The mill can be brought into play now that the dirt and dust that was mixed in with the barley has been discarded. Two steel or stone fluted rolls should be set in motion, so that the milling process steadily takes place and is repeated, with the result that the grain is now thoroughly milled.

Compared with the milling of green malted barley, the milling of malted barley is easy, since the grain has been dried. However, care must be taken that the rotation of the fluted rolls does not cause fire through friction. In this way, the first stage of malt production is complete. The milled malt is packed into sacks of fixed capacity and these are then taken to the mashing room as required. Normally the storeroom is adjacent to the mashing room and set one floor above so that the work is made as easy as possible and can proceed apace.

The dried malt barley does not improve through storage and so it should be sent on to be mashed as soon as possible. However sometimes grain is unobtainable, and so it is advisable to keep a slight surplus in store, and that the storage facilities be constructed in such a way as to take this need into account.

*Plant for drying of draff
(Dryers = 5
 Boilers = 2)*

Mashing

1. The mashing schedule

The production of wort takes place only on the Monday of a given week. It is then fermented on Tuesday and Wednesday. On Thursday, Friday and Saturday it is distilled. On Sundays, apart from a few specialist workmen who remain to take care of the malt, the distillery is closed. If the mashing were carried out on a Wednesday, it would be impossible to complete the fermentation and distilling processes by Saturday. Occasionally, production of wort cannot be completed on Monday, but, for the most part, the process is always over by noon on Tuesday. In continuous distilling, the wort is produced on Mondays, Tuesdays and Wednesdays. It is then fermented and the distilling process completed during the night on Saturday. However, here I am reporting on pot still whisky, and so, only after the distilling has been carried out twice or even three times, can the final product be obtained. Therefore, it is perfectly reasonable that the three days of Thursday, Friday and Saturday are required for distilling. In the event that distilling cannot be completed on Saturday, the regulations require that the still is then sealed in the presence of the Excise Inspector and opened up again on Monday from 1.00 a.m. For that reason, the wort is only produced on Mondays and the distilling is carried out day and night from Thursday to Saturday only. Before describing the production of the wort, it will be necessary for me to outline the structure of the mash tun.

2. The device used for producing wort and its apparatus

The mash tun is made entirely of iron and it normally has a capacity of 300 *koku*. In large distilleries, there are huge mash tuns that have a capacity of 525 *koku*. As for the 300 *koku* type, the diameter is 18 *shaku* and the height is 7 *shaku*. The tun is circular. In distilleries such as our own home distillery in Japan, the raw material for the wort is sent under extreme pressure to the mash tun from a steamer, and so the mash tun needs to have a lid. However, in the case of pot still whisky, the raw material is malted barley and so the mash tun is unlidded.

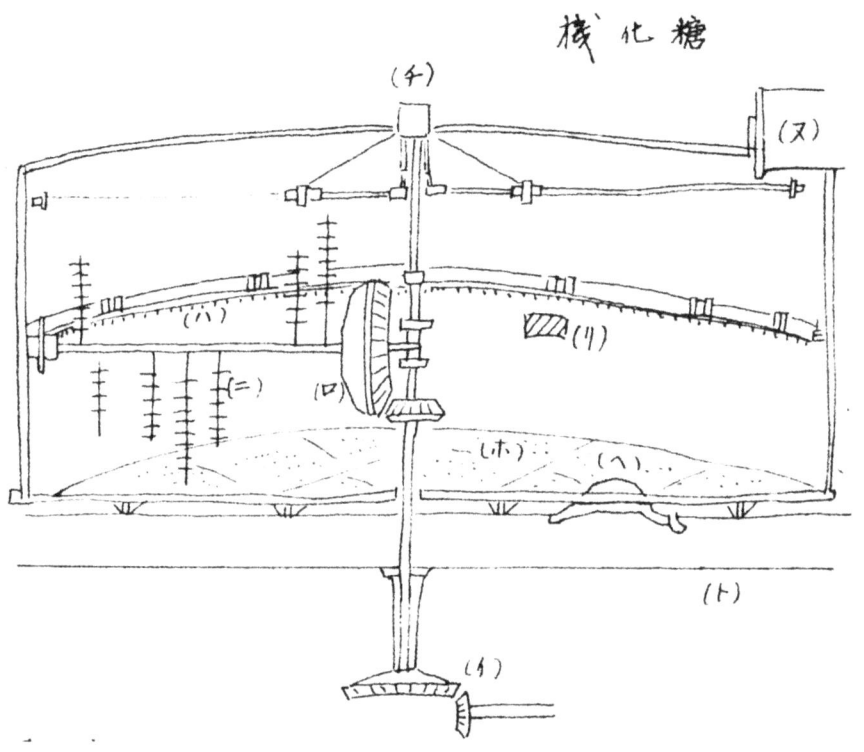

Cross-section of a mash tun

(ヌ) refers to the Steel's masher, shown in detail on the following page.

Explanation of the Diagram:

Movement is supplied from the cogs (イ) to the axle (ロ) and thence to the teeth (ハ) cut into the perimeter of the axle, and steadily the interior workings revolve so that the wort is thoroughly mixed. Differently from the mash tuns used in an alcohol distillery (i.e. Japanese distillery), there are no mixers.[9] The reason for this lies in the fact that not all of the mash will turn into wort – only a part of the strained liquid will be sent to the wash backs and so the mash is extremely thick in consistency. A great deal of power is required to mix it, carrying as it does a lot of husks within it. The task could simply not be done using mixers and so, as can be seen from the diagram, a strong rake is used. The wort is strained through the countless small holes in the base (ホ). The base consists of numerous iron screens and is constructed in such a way that these can be removed at the end of the process for cleaning. The base faces the outlet (ヘ) and is set at an incline, and the wort proceeds through this outlet to the pipe (ト). It is then sent to the cooler by means of a pump. (チ) shows the part that is there merely to support the axle, while in some types of mash tun they have installed moving cogs in the upper part.

23

(リ) indicates the apertures that allow the material left in the sieve to be ejected. They measure 2 *shaku* by 2 *shaku* 5 *sun*. During the mashing process, the tun is completely sealed from the outside so that nothing can drip out. Two or three workmen go into the mash tun after it has done its job. Using spades, they shovel the remaining residue through this outlet out onto a waiting horse cart. It is then taken to the place where the residue will be treated. More will be said about disposal of residue (draff) in a later chapter.

There are no coils installed in the mash tun for the purpose of heating or cooling. The reason for this is that the raw material is powdered malt barley and so merely by pouring in mashing liquid that has reached the prescribed temperature, the process can take place. The technicians take great pains to ensure that saccharisation can be achieved by carefully measuring the temperature of the water to be added.

The diagram shows the device used for mixing the malted barley and water. It is cylindrical in shape, being 2 *shaku* 5 *sun* in diameter and 6 *shaku* in length. There is a single axle inside, connected to an iron mixer that rotates 150 times per minute by means of a wheel located on the other side. In this way, the milled grain is thoroughly mixed with water before being sent into the mash tun. There is no clumping and the production of wort is achieved to perfection.

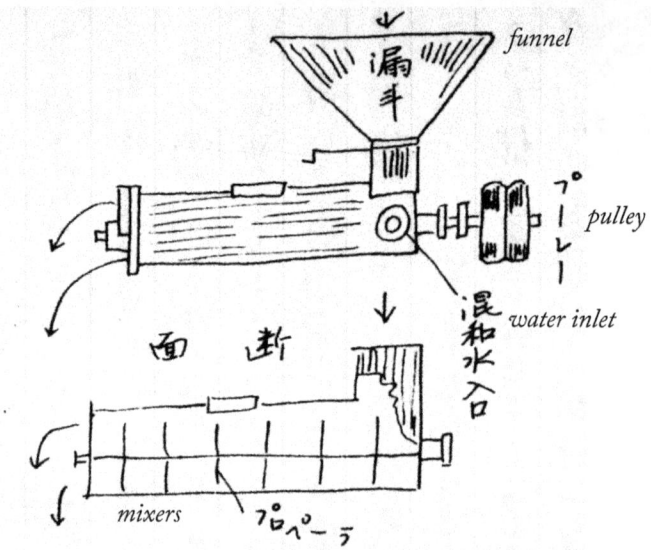

Cross-section of device used to mix malted barley and water. (Steel's masher)

3. The effect of mashing

The proportion of malted barley to water varies somewhat from factory to factory, just as is the case in Japan when it comes to the production of *sake*. Therefore, what follows is an account of the most representative of the various processes, the one that takes place over the period of a week, when 200 *koku* of malted barley is used on Monday to produce the wort. In large-scale operations, the amount used might reach 700 *koku*. At Hazelburn, the average is 400 *koku* but the *method* of production does not alter, and so I will confine myself to an account of how 200 *koku* are used to produce wort over a period of one week.

The pot still method uses only dried malted barley and differs from the continuous method in not mixing in any other type of milled grain.

In the UK, the unit of measurement is the bushel for grain (2 *to* 1 *gō*) and the gallon (2 *shō* 5 *gō* 2 *shaku*) for water, but I have here converted all the measurements into Japanese ones.

The sequence of mashing that I experienced first-hand was as follows:

Mash no.	Date	Time	Volume of malted barley (koku)
1	Mar 1 Mon	05.00	60
2	"	13.00	55
3	"	21.00	50
4	Mar 2 Tue	04.00	35
Total			200

Before adding the malted barley, some water has to be spread over the inside of the tun. This is so that the countless small holes on the base of the mash tun do not become blocked and so that the warm water will warm the tun and hasten the production process.

Mash 1 requires 60 *koku* of malted barley to 100 *koku* of water (this malted barley has a bulk density of 216 *monme* per *shō*). The water temperature needs to be controlled in order to heat the wort gradually from 50°C to 57° C. The final temperature reached must be 61 degrees C. During this time the mixer is turning ceaselessly. When the temperature of the wort reaches 61 degrees, the mixer is stopped and the mashing begins. After an hour and a half, straining begins and the wort is sent to the cooler. However, before this process comes to an end, the pipe on the bottom is closed and once again hot water is added to the mash tun and wort production recommences. This is so that some of the first batch of wort remains in the vessel and the enzymes contained within it are made available in order to complete the mashing of the next batch.

For the second stage of mashing – 60 *koku* of water at 77 degrees is added. Then mixing takes place in order to bring the temperature of the wort to 65°. After an hour of mashing, the wort is strained. 60 *koku* of water is added to an initial 100 *koku* and so a total of a 160 *koku* is used. However once it has passed through the perforations at the bottom of the mash tun, only 135 *koku* will reach the wash back, the remaining 25 *koku* being absorbed by the draff. Water is then added once again to cleanse the draff. This liquid is not sent to the wash back. This is because if you use the same raw material three times for mashing it becomes very thin – registering less than a 3 on the Balling Scale - so instead it is sent to a heated iron tank. It is used instead of water for the second mash. For the 3rd cleaning stage, water at 87 degrees is added and thoroughly mixed. Great efforts are made to ensure that all sugar is washed out of the residue. When the cleansing is finished, the workmen immediately take up their shovels and remove the draff. Eight hours are required to elapse from the first adding of the grain to the removal of the draff.

Mash 2 is started at 1.00 in the afternoon and it finishes at 9.00 in the evening. The amount of malted barley used is 55 *koku* – 5 *koku* less than for the first mash. The amount of water used is the same as that used in Mash 1, i.e the amount that can send a total of 135 *koku* of wort to the wash back. Therefore, as the amount of grain is less than that used for the first mash, the wort is less sweet. The process for Mash 2 is the same as that for Mash 1, except that, in place of water, the third stage thin liquor is used.[10] This liquor is heated in the iron tank to a temperature of 75 degrees. It is skimmed, and the clear liquid sent via pipes to the mash tun.

After the 2nd mashing stage, 160 *koku* of water heated to 87 degrees is added once more for the 3rd stage cleansing of the tun, and, at the same time as mashing takes place, washing is also done and, after one hour, just as was the case with Mash 1, the liquor does not go to the wash back but rather to the heated tank, where it is used for the next mashing. In this way Mashes 3 and 4 are carried out in sequence and, despite the fact that for Mash 4 the amount of raw material is at its lowest, the amount obtained is still 135 *koku* as before.

In this case, for Mash 4, the third batch of cleaning liquid is not sent to the heated tank but is sent just as it is to the wash back. Therefore, the amount of water required for Mash 4 is different from before in being 80 *koku* for the 1st mashing stage, 40 *koku* for the 2nd, and 40 *koku* for the 3rd. When it is sent to the wash back, it has acquired a total volume of 135 *koku*. In the case of the final mash, it is not possible for mashing liquid to be sent to the heated tank and used for the next mash – the next mash will not take place until the following Monday and it is not possible to preserve the liquid until then.

The liquor from Mash 1 is sent to wash back No. 1, that from Mash 2 to wash back No. 2, and in this way Mash 4 will be sent to the No 4 wash back. For this reason, fermentation proceeds in that order and distilling can then begin. The process has been set out in the table opposite.

4. The draff

As previously stated, Scotland is a country where livestock farming plays an important part and in the winter, owing to shortages of feed, most farms buy up draff. These farms buy the draff at 2.50 yen per *koku* and the owners come directly into the distillery in horse-drawn carts and take the draff away while it is still fresh. If the wet draff is shovelled up out of the mash tun as it is, analysis reveals that the water content is 78% while the starch and sugar content is 10.5%.

There are also proteins and fats, etc. present. If we take 200 *koku* of raw material, and mash it, then sometimes the amount of draff obtained will amount to as much as 225 *koku*, and so there is no way that it can all be carted off to the farms. In summer especially the hills are everywhere covered in an abundance of green grass, and considerably less draff is consumed. In Campbeltown, a drying plant for draff has been established in the town. Each distillery sells its draff to that plant, and that is where the drying is carried out. There is no need for each distillery to have its own drying facilities on site. The mechanism for drying is described in detail in the report on the subject of column distilling of grain whisky that I submitted last year. As I set out in that report, numerous steam pipes are fitted within a cylindrical vessel shaped like a boiler. The cylinder is made to revolve, sending the damp draff from the opening on one side to the place from where it is discharged on the other, thus drying it in the process. In places where continuous distilling is carried out, each distillery

Table of the mashing sequence

1st Mash	60 koku of malted barley	Temp	
Stage 1	100 koku of water	62°	135 koku of wort (Wash Back No. 1) *Av. On Balling Scale = 29*
Stage 2	60 koku of water	68°	
Final stage	160 koku of water	73°	Sent to heated tank to be used for 2nd mash
Removal of draff			
2nd Mash	55 koku of malted barley	Temp	
Stage 1	100 koku liquor	62°	135 koku of wort
Stage 2	60 koku liquor	68°	(Wash Back No. 2) *Av. On Balling Scale = 29*
Final stage	160 koku liquor	73°	Sent to heated tank to be used for 3rd mash
Removal of draff			
3rd Mash	50 koku of malted barley	Temp	
Stage 1	100 koku of liquor	62°	135 koku of wort
Stage 2	60 koku of liquor	68°	(Wash Back No. 3) *Av. On Balling Scale = 27*
Final stage	160 koku of liquor	73°	Sent to heated tank to be used for next mash
Removal of draff			
4th Mash	35 koku of malted barley	Temp	
Stage 1	80 koku of liquor	62°	135 koku of wort
Stage 2	40 koku of liquor	68°	(Wash Back No. 4) *Av. On Balling Scale = 23*
Final stage	40 koku of liquor)	73°	

will have its own dryer. This process requires large quantities of steam and so, in the case of pot still distilleries, which do not even have a boiler installed, the process has to be carried out on a cooperative basis, and it seems only natural to have a special factory set up for this purpose.

This machine, if further developed, would be very useful for our Head Office, as it can be used not just for dealing with draff, but also the lees produced by distilling other types of alcohol, such as *mirin*[11], or the *shōchū* distilled from *sake* lees at New Year. In this way drying can be done without the need for vast swathes of vacant land and a large workforce, as would be the case when drying under the sun. At times when one has no choice but to buy in large quantities of raw materials with a high water content for the purpose of distilling, if only one had a machine like this, drying of the grain could commence as soon as it comes in, prior to its being stored.

However, this is not to say that it is absolutely necessary to have one of these machines, merely that it is by far more convenient to have one at one's disposal. When attempting to experiment in Scotch whisky distilling on an initially small scale, it will be a secondary consideration. In Japan, livestock farming is not widespread. I have given some thought as to how we might dispose of the draff and, fortunately, have managed to come up with a proposal, which is that it could be used in place of bran when making rice malt. Not only does it differ little in its composition, but, when properly strained, the moisture content of the lees also corresponds very closely to that of *kōji* (rice malt) spores. The draff would be cooled to an appropriate temperature once it has been removed from the mash tun. Then it would be taken to a place where rice malt is being produced, there to be used instead of bran. In fact, I believe it would prove preferable to bran, from the point of view of utilising material that would otherwise be discarded. However, unless we carry out a trial, it is impossible to say definitively whether this would be a viable option.

I have now completed my notes on the mashing process and disposal of the draff as it obtains here in Campbeltown. What follows is an account of my thoughts on the question of which aspects of the process as it is carried out in Scotland should be improved if we are to begin mashing in Japan.

What happens in Scotland is that, for some reason, water for mashing is supplied in equal quantities. It is only the quantity of raw material that is then gradually reduced with each mash. This is because distilling needs to be completed by Saturday afternoon so that the workers can spend Sunday peacefully at their leisure. The fourth mash is as thin as possible in order that the distilling may be completed in as short a time as possible.

If the fourth mash were the same as the first, that is to say – using 60 *koku* of malted barley, with 135 *koku* of wort produced - then a lengthier period would be required for distilling and so it could not be completed by Saturday afternoon. As to the question of whether all the distilleries in Scotland work to this schedule, there are places where the proportion of raw material for the fourth mash is the same as for the first, but in those cases the process must begin four hours earlier than is the usual case, that is to say 1 a.m. on Monday morning. I would support adopting the latter process, i.e. that which uses the same proportions for all the mashes.

Which process will be used is a matter that should be decided on the basis of the workmen's hours of employment, and this matter is at present undergoing a period of rapid change. I see problems in regard to resolving this question as it relates to Japan. If we follow the convention of allowing two public holidays a month as has hitherto been the case, mashing does not have to commence at one o'clock in the morning as it does in Scotland; it can start at the normal time at which the working day starts in Japan. Mashing could be then carried out sequentially. In my opinion, the mash tun should be constructed in exactly the same way as in the UK. However, its capacity should be smaller. It should have a diameter of 12 *shaku* and a height of *6 shaku*, giving a capacity of approximately 100 *koku*. That would be sufficient in my opinion. A mixer for raw materials must be attached.

The mixer ought to be adapted so that the toothed wheels are affixed to and rotate at the top of the tun. Mashing should be done as it is in Scotland, twice daily with a three-step cleaning process accompanying each step, each mash taking seven hours to complete. The amount of malt used will be 30 *koku* to give 70 *koku* of wort, after which the first fermentation will begin. For that reason, a total of 60 *koku* of malted barley will be used per day and the total amount of wort produced will be 140 *koku*. The work will begin at 7 a.m. and stop at 8 p.m. on the same day.

As for Scottish distilling methods, these will be described in the chapter under that title but here I have to record the fact that the mashing process will be influenced by whether there is only one still or whether there are two. There are no Scottish distilleries which have only one still.

Usually there are at least two – one for the first distilling and one which is designated as the spirit still. My recommendation, however, is for us in the beginning to distil all our wort in one still and then, in order to separate the spirit to be used for whisky and other high-quality spirits, for us to conduct a second distillation using that same still. Since mashing cannot be done every day, there will be a need for such a separation. The first to fourth mashes would be done over a two-day period in batches and, when the fourth mash is completed, then mashing would be temporarily suspended so that distilling may be carried out.

Therefore, as compared with patent still manufacture, pot still manufacture is a somewhat tortuous process, but it is important to realise that it is only through patience that one may produce a spirit of superb quality. I will report further, summarising my views as to which strategies we should employ after I have completed my report on distilling. Here I will confine myself to setting out my view of the mashing process.

Cooling

Before the mashing liquid is sent to the cooler, it is first sent to the underback. The underback is situated at a height above the cooler and is placed in such a way as to facilitate the flow of liquid down into the cooler. The solids that are present in the liquid sink to the bottom of the underback so that when the liquid runs into the tank it is almost completely clear.

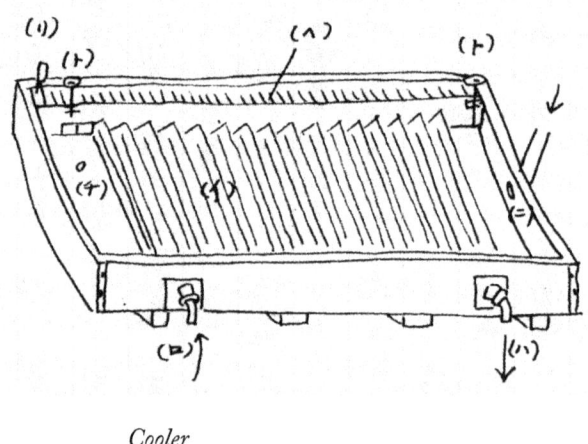

Cooler

The cooler is as shown in the diagram.
The mashing liquid is delivered through the pipe on the right (ニ) and then goes through the copper plate (イ) and in this way is cooled. The water required for the cooling process enters through the pipe indicated by the symbol (ロ) and, passing through the copper plate, is expelled through the pipe indicated by the symbol (ハ). The cooled liquid makes its way to the washback through the aperture (チ) into a pipe. As for the structure of the metal plate (イ), the following is a detailed description of same. As you can see from the next diagram, two types of copper tubes are arranged in alternate sequence, and it is through these tubes that the cooling liquid passes (shown by a note underneath the diagram).

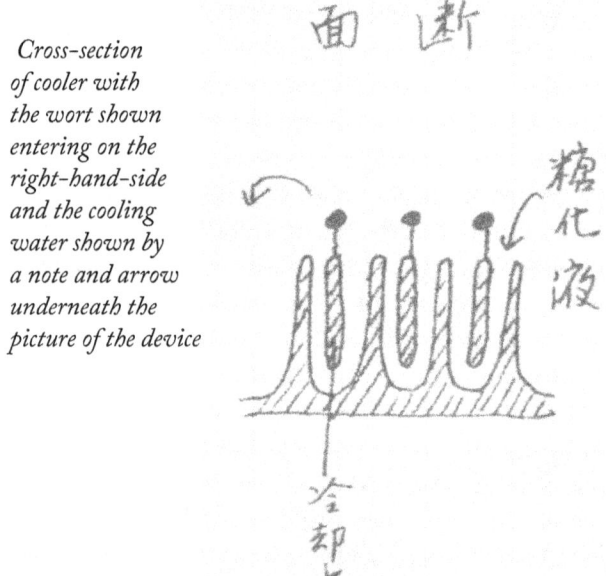

Cross-section of cooler with the wort shown entering on the right-hand-side and the cooling water shown by a note and arrow underneath the picture of the device

The wort (shown entering on the right-hand side) moves constantly up and down and from right to left and is cooled. The connecting of the tubes is at the side(s) of the cooler. Since an incline is built into the device, the liquid flows along quite naturally.

In order to clean the cooler after it has finished its work, the handle (リ) is pulled, opening the holes drilled into the plate (ホ) and the various materials lying in the wells (イ) are flushed into the space indicated by the symbol (ヘ). Now, assuming that there are 40 wells (イ) that means that, when the handle (リ) is pulled, all 40 open at the same time. (ト) simply refers to the stopper – when this is opened, the various solids in the space indicated by the symbol (ヘ) are expelled through the pipe indicated by the symbol (チ). This pipe (チ) is divided at its bottom end in such a way as to separate that which will be sent to the wash backs and that which will be discarded.

At the entrance and exit of the cooler, a pipe is set at a height of roughly 2 *ken*, so that, in the unlikely event of a severe increase in water pressure, water will be blown up and out through the pipe, thus averting any breakage of the copper tubes (イ).

Since in this kind of cooler the mashing liquid cools as a result of constant contact with the outside air, there is a danger that bacteria may enter during cooling. However, viewed from a different angle, one must remember that cooling is only possible through contact with the outside air. Moreover, the length of time during which cooling is carried out is extremely short and so excellent outcomes are achieved. Each time a mash is completed, the workers clean the cooler, and so hygiene is maintained at all times.

The effectiveness of the cooling process is extremely good, and the smallest cooler can cool 45 *koku* of mashing liquid in an hour. The cost is 400 yen at present. A large cooler can cool 900 *koku* per hour and costs 5,000 yen. Of course, the effectiveness of the cooler depends on the temperature of the water used. A temperature of 15 degrees C has been taken to obtain the figures cited above. The majority of Scottish distilleries make use of this kind of cooler, but at Gartloch in Glasgow a vertical cooler is used. This is exactly the same in composition as the one described above, except that rather than being horizontal, it is vertical. The vertical cooler is more in contact with the outside air on account of its larger surface area, and so its cooling function is better than that of a horizontal cooler. There is no need to devote space to a description of how the cooler functions, but it is important to remember that the temperature of the mashing liquid at the outlet should be constantly monitored and the amount of wort being sent through it adjusted accordingly. The cooling temperature of the wort is 21 degrees C. I would like to see us experimenting with this type of cooling device in Japan. If we could make a cooling device of 4 *shaku* in width, 1 *shaku* in height and 7 *shaku* in length, it should be possible to cool 40 *koku* an hour. It would be cheaper to have it manufactured by a manufacturer of copper goods in Japan than to order one from the UK. We will not just require a cooler, but also mash tuns and stills and so on. Much of what we need will have to be brand new. I suggest that this all be discussed after my return to Japan.

Fermentation

The wash back is almost always of the same shape throughout Scotland, i.e. cylindrical. It is made from *kashi* (Japanese Emperor oak) or pine.[12] It can be used for a period in excess of ten years. In grain distilleries, one finds pipes introducing compressed air, heating and cooling pipes, and yeast gatherers attached to the vessels, but, in the case of pot still distilling, there is nothing extra in the way of equipment inside the vessel – only the two wooden propellers that turn steadily in order to prevent bubbles frothing up and escaping during the fermentation process. The capacity of these vessels is generally uniform – in large distilleries there are just more of them.

At Hazelburn, there are eight wash backs with a capacity of 150 *koku* each. Their diameter increases towards the bottom, being 10 *shaku* in the middle. The height is 14 *shaku*. The wash backs have a wooden lid into which a window has been set so that the fermentation process may be checked and observed, and samples drawn so that the sugar level may be measured. A pipe of 4 *sun* in diameter leads into the bottom of the vessel and the cooled wort flows in through this pipe. After fermentation this same pipe is used to move the wash to the wash charger below.

As for the yeast, some of the wort is drawn off and yeast is dissolved in it at a ratio of 10 *koku* of wort to 800 *monme* of compressed yeast.[13] The compressed yeast is obtained from companies producing beer or stout (a special kind of beer produced in the UK). The sediment from the beer or stout-making process is compressed and put into sacks for sale. This sediment is almost the same in form as the *sake* lees used to make Nara pickles.[14] The quality is the same as that of the yeast compressed at continuous distilling plants, but that kind of yeast is only used in bread-making, not in the fermenting process for pot still distilling.

The yeast obtained from companies making beer is bought at a rate of one yen for 1 *kan* 600 *monme*. This is cheap and, since it can be obtained as and when necessary, pot still distilleries – which do not carry out grain distilling – do not cultivate it themselves. However, in Japan, we have not reached the stage of being able to gather pressed yeast of reliable quality in our beer manufacturing. Therefore, as our company already has a laboratory of a considerably high standard, I think we should try our hand at producing pure cultures ourselves.

Tests to be carried out during fermentation are simply the measuring of the sugar content to see to what extent it has decreased. The device used for measuring the amount of sugar used in the UK was invented by Bate and is called Bate's Saccharometer. Since it is made entirely of brass, there is no worry that it might get broken.

Now if one places the instrument into distilled water at 15 degrees C, it stops at point 0 at the top of the scale. Gradually, as the specific gravity increases, it floats until the 30 degree mark at the bottom of the scale is reached and then the first weight is attached at point (イ) at the base of the saccharometer. When 60 degrees + is reached, the second weight is used, and in this way it is possible to calculate measurements up to 120 degrees. This is mostly used in the brewing of beer and it is extremely useful for that purpose. It allows the brewer to determine right away how many pounds of extract are contained in a barrel containing 35 gallons.

Supposing that the meter is showing 15 degrees, then

15 x .36 = 5.4

That is to say, in a 35 gallon barrel there is contained 5lb + .4 of extract. To determine the usual specific gravity using the saccharometer, the following formula is used:

Specific gravity = 1 + 0.002 $^7/_9$ x number shown on the scale

Supposing the scale shows 50:
Specific gravity = 1 + 0.002 $^7/_9$ x 50 = 1.138

In Scotland, none of the values pertaining to Specific Gravity are expressed and so these formulae are unfamiliar to a foreigner such as myself, but they did form part of Professor Tsuboi's lectures.[15] It is only after coming here and seeing it all with my own eyes that I have been able truly to appreciate how valuable his lectures were. However, one has to acknowledge that they are manufacturing whisky in a manner that is, for the most part, diametrically opposite to that which we were taught at university.

In this report, all the saccharisation values expressed on the Balling Scale are derived from the Specific Gravity values obtained by means of Bate's Saccharometer, then calculated using the afore-mentioned formula to obtain the Balling value. Next, if one shows the reduction in the degree of saccharisation in tabular form — as can be seen below — it may be seen to be more or less consistent with the time elapsed and that the process of fermentation is not determined by a fixed level of alcohol present. Furthermore there

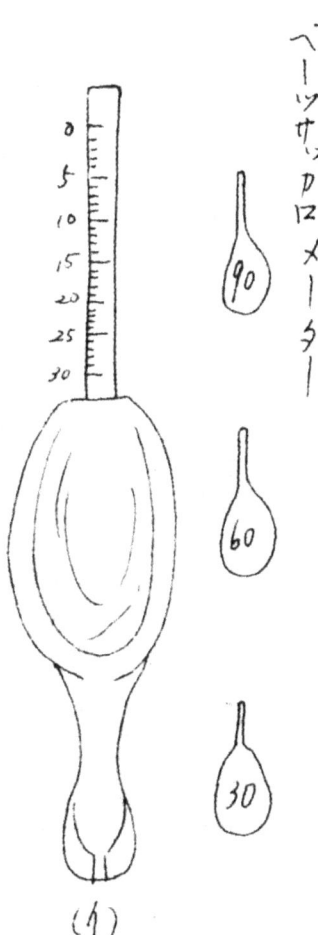

Bate's Saccharometer

is no need for the use of a microscope. The degree of saccharisation may be determined through the use of one simple saccharometer.

Fermentation times
No 1: 48 hours
No 2: 47 hours
No 3: 44 hours
No 4: 40 hours

By decreasing the raw material used in the fermentation, the time required may be reduced.

About 3-4 hours after fermentation commences, a violent bubbling arises and, at that stage, a belt is affixed to the propeller at the top of the vessel and it is set revolving so as to disperse the bubbles. With the passage of time, the liquid might overflow and, in order to prevent this, soap may be added – but this is hardly ever done.[16]

During fermentation, the wash is not mixed. Rather, it appears that fermentation is left to take place naturally and the result is in no way adversely affected. In pot still distilleries, just as in grain distilleries, there is no pH adjustment of the wash. After fermentation, the alcohol content is at somewhere between six and seven percent, but no level is set by the Excise Department, nor does it see any need for same. However, speaking as a technician, I myself feel that it is necessary to determine the exact level of alcohol present in the wash prior to distillation.

From time to time, the Exciseman will test the volume of wash. This is made difficult by the bubbling, as it is then difficult to know how much of the vessel is free space. For that reason, there is a cork of 1 *shaku* square attached to one end of the measuring stick. It is put into the vessel and set afloat so as to determine the level of the liquid and the amount of free space.

After the wash is fermented, no acidity will be detected if one tastes it. However, a slight taste of peat may be left on the tip of the tongue. At this stage, it will taste more like whisky than wash. This taste was completely undetectable before when the liquid was wort; it only appears after fermentation. This is a truly interesting phenomenon.

In pot still distilling the yeast is not harvested and so, once fermentation is over, the wash is immediately sent to the wash charger and distilling commences right away. However, prior to distilling, the wash is left to rest for about two hours so that all the sediments settle out and only then is it sent to the still. Cleaning the wash back: the wash back is disinfected as soon as the wash has been sent to the still. Lime is then painted on the inside of the vessel and left for a day. The vessel is then cleaned once more. It is washed before and after with boiling water, then disinfected by steam. Lime is painted on once again and after that the wash back may again be used for fermentation.

As to how fermentation may be carried out at Head Office in Japan, firstly two wash backs should be constructed from oak[17], and propellers affixed at the top for the purpose of dispersing the bubbles. The vessels should be 7 *shaku* in central diameter and 13 *shaku* tall, with a capacity of 80 *koku*. In this way, we would be able to ferment 140 *koku* of wash at the same time using the two wash backs. As previously noted, the wash backs would probably not be for summer use. However, once we begin our experiments in making

Scotch whisky, there is no need to insist on obtaining barley in summer, or water when it is in short supply; we can certainly commence at the beginning of December and end in March. That should be the best time of year for production to take place. In that way we will be working during the period when the climate in Japan most resembles that of Scotland, and I believe that that is the safest way of ensuring the best results.

As previously stated, what we need to work towards is ensuring that the temperature of fermentation does not exceed 33 degrees and that there is as little acidity in the wash as possible. We also need to decide upon how to get rid of the draff.

Distilling

There are two types of distillery: those in which distillation is carried out twice and those where it is done three times.[18] The former type produces distilled spirit with an alcohol content of 60 degrees and the latter 82 degrees. The Campbeltown distilleries all distil twice. All the alcohol is distilled from the first wash, and so the alcohol content of the resulting liquid (low wines) is, on average, between 15 and 23 degrees (abv). Naturally, at this stage, it is a cloudy white liquid that is unsuitable for drinking. For convenience, here I shall call the still used for the first distillation as "No. 1", and that used for second distillation "No. 2".

The first distilled liquor is distilled once again by means of the No. 2 still. The fore-shot, whisky and tails are collected in separate receptacles. (Normally, the fore-shots and the tails are collected in the same receptacle.) The fore-shot and the tails are brought together at the time of the second distillation. The whisky has an alcohol content of 63 degrees, the fore-shot 65 – 70 degrees, while the tails have an average alcohol content of 25 degrees. The spent lees have an alcohol content of 0.2, but I believe this loss is unavoidable.

How Excise is applied in the UK

- The capacity of every still must be measured and equipped with a device that allows a sample to be drawn for testing.
- On top of each still a support must be fixed and made stable and all the pipes, including the flue from the furnace, must be safely secured.
- The still must be equipped with a safety valve and an air valve.
- The outlet for the whisky which has passed through the worm condenser (used for cooling) should be encased in a glass box and a device which seals the box must be attached.
- On the bottom of the still there should be a pipe through which the spent lees and pot ale are expelled, controlled by a stopcock, which it is necessary to attach within 3 *shaku* of the base of the kiln. (The technicians at the distilleries are calling for this article to be removed from the regulations. This is because the requirement is that the stopcock must be closed off as well. If the distilling has finished, then not only do the manhole and the air valve have to be sealed, the opening to the flue has to be sealed too. For some reason, the Customs and Excise requires that the discharge stopcock must be sealed as well, and this has given rise to complaints from the distillers.)
- In distilleries where no receptacle for the pot ale is connected to the still, it is not possible to obtain permission from C&E to open the stopcock within six hours of the start of distilling. (The reason for this is to

prevent the discharge of any liquid that has not been fully distilled.)
- Apart from what has been described above, it is not permitted for any other piping to be attached to the still by means of any further openings.
- It is essential that the low wine receiver used in the first distillation is separate from the feints receiver used for the next distillation.
- Once whisky has been sent to the spirit receiver, no further distillation takes place, but the need to gain the permission of C & E does not cease here.
- Only upon having received the permission of the C & E, or having received said permission on behalf of the Exciseman in the case of his being absent on a business trip, the spirit receiver may be opened and the quality of the spirit tested.
- Between 11 p.m. on Saturday night and 1 a.m. on Monday morning, it is forbidden to use the still.
- Prior to all the wash having been transferred from the wash backs to the wash chargers no distilling can be commenced, nor can it begin until two hours have elapsed since the transfer was completed.
- The volume of the still must be at least 400 gallons (approx. 10 *koku*)

On the basis of the regulations set out above, the shape of the still is almost completely uniform wherever one goes and the following diagram shows a typical example.

As shown in the diagram on the next page, it is a simple helmet-style shape. The minimum capacity is 10 *koku*, going up to a maximum of 300 *koku*. The normal volume will be 100 *koku*. The still is made entirely of copper and, if one takes it to have a capacity of 100 *koku*, at its widest part it will have a diameter of 8 *shaku*. It is 12 *shaku* from the neck to the bend at the highest point of the still, where the diameter is 3 *shaku*.

The low wines and feints inlet, manhole and air valve are set three quarters of the way up the body of the still. The tank (I have chosen to call the round part at the base of the still the 'tank'), is filled with either wash or low wines either to 2/3 or ½ of the total capacity. This is because, once distilling commences, the volume of liquid will expand because of the heat, and the bubbles created might overflow if it were filled any higher than that. The overflow of bubbles can affect the quality of whisky being produced.

As shown in the diagram, the operation of the air valve is controlled by rotation of a stopcock.[19.] As soon as distilling has ceased, it is opened up and the internal pressure released. It is not unknown for this step to be neglected, and injury be caused to the workers when the manhole is opened right away.

After the first distillation has been completed and prior to the wash being sent for second distillation, this air valve must be opened. If this step were to be forgotten, then cold wash will suddenly flow into the hot still, creating a vacuum that might end up damaging the still. Care must therefore be taken at all costs.

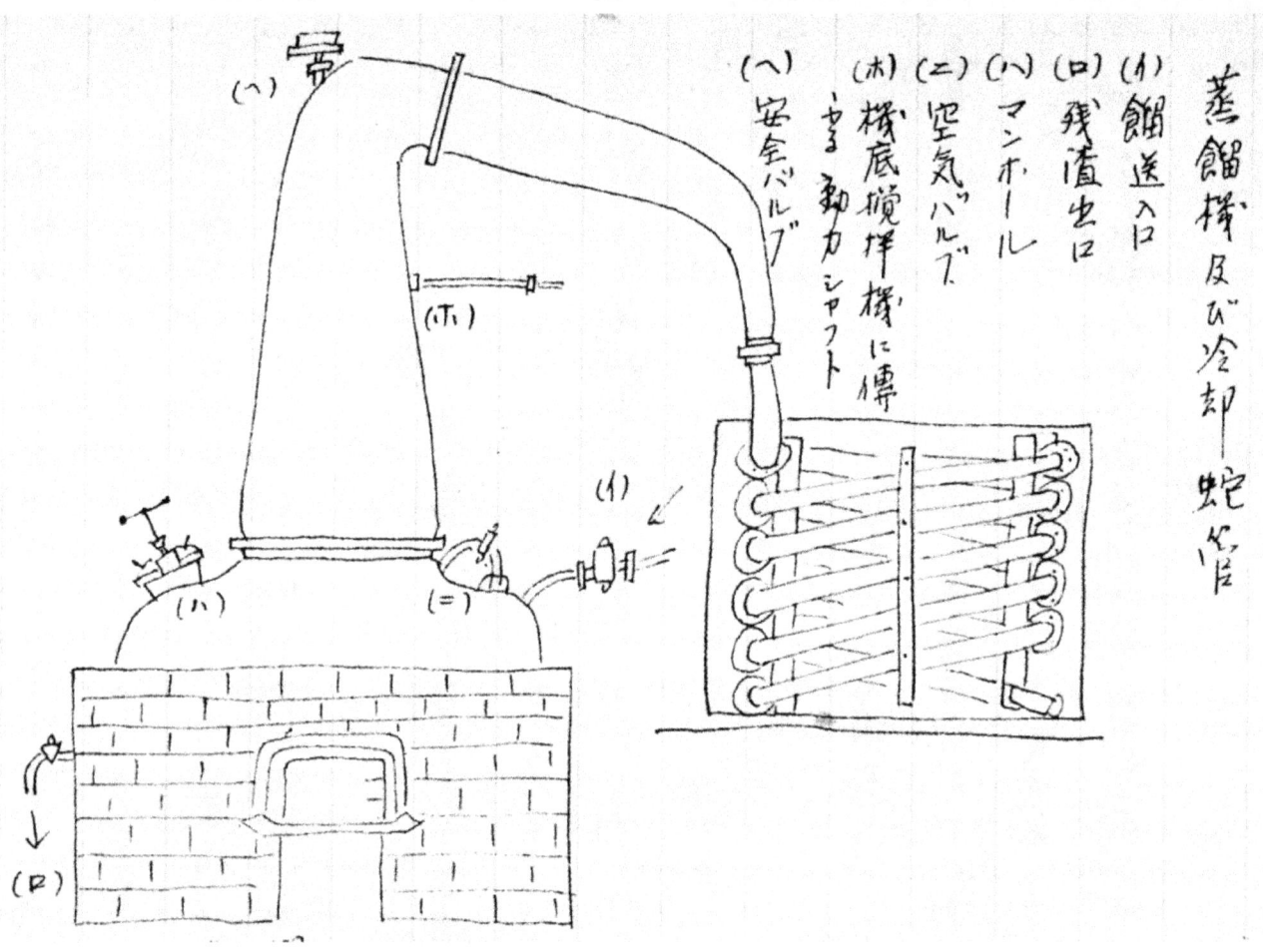

The Still and Worm Condenser
Key to diagram:
(イ) *Filling inlet*
(ロ) *Lees discharge*
(ハ) *Manhole*
(ニ) *Air valve*
(ホ) *Shaft driving the mixer*
(ヘ) *Safety valve*

(ホ) indicates the position of the shaft that drives the mixer at the bottom of the still. It is only present in still No. 1, not still No.2, as the second distillation is simply a repetition of the first, and so mixing is not necessary. The mixing apparatus is controlled from outside by means of the rotation of a single axle. A multitude of chains gradually mix the wash at the bottom of the still. Power for the movement of the mixer is supplied by a water wheel that utilises the water outflow from the worm tubs.

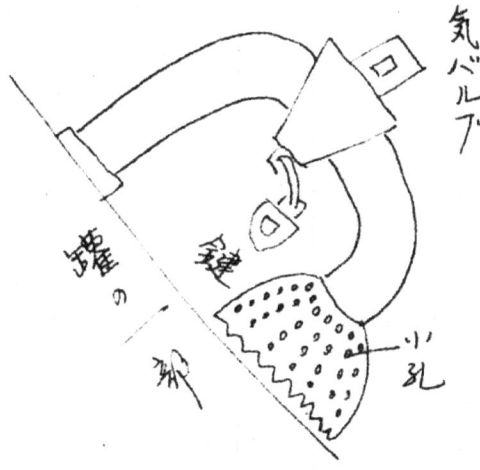

Enlarged drawing of air valve. Key:
空気バルブ *air valve;*
鍵 *key/padlock;*
小孔 *small holes*

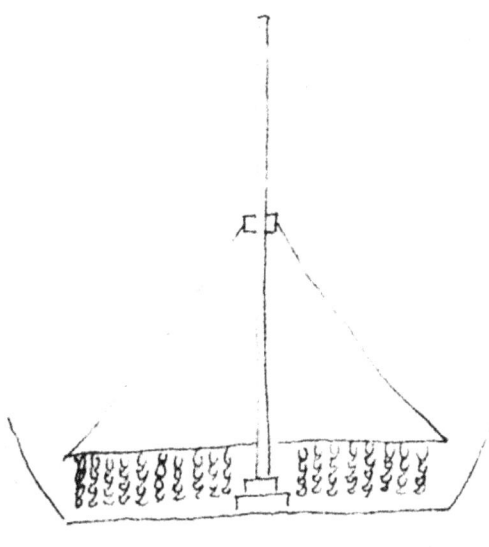

Mixing Apparatus at the bottom of the still [The Rummager]

The cooling method involves the use of a worm, the diameter of which narrows towards the bottom. At the top it is 1 *shaku* and at the bottom 5 *sun*. It has 6-7 levels. The vessel that holds the cooling water is made out of oak and has a diameter of 10 *shaku* and is 13 *shaku* high.

In all pot still distilleries, the cooling apparatus is attached outside the still room, and so most of them are at the mercy of the elements. The pipe at the head of the still has to be led out to the outside by knocking through the wall of the room housing the still, and this tends to look rather unsightly. However, from the point of view of cooling, it is very effective. Moreover, it allows full use of the room for distilling. In the case of distilleries that have 3-4 stills, it is therefore not necessary to widen the building if one's aim is simply to accommodate the worm tubs. It is perfectly understandable that they choose to do it in this way.

The photograph shows the worm tubs that are located to the west of the Hazelburn Distillery. The discharge flows into a clear stream that runs underneath and, on the rare spring days that they have in Scotland, children play in the water. Of all the photographs that I have taken in Scotland, it seems to me that the photographs I have taken of that scene (i.e. the children playing in the burn) are the freshest and, aesthetically speaking, the most successful.

Scottish winters are wet and liable to be overcast and so the climate does not lend itself to photography. From the beginning of spring and into the summer, fine days follow one after the other, but inside the mash room, for example, it is dark and so it is impossible to take any photographs in there. I regret very much that this has meant that I have been unable to include interior shots in this report.

The photograph on the next page shows the worm tubs situated to the east of the Hazelburn Distillery, together with the roof of the drying room for malted barley. In Campbeltown, the roofs of the buildings run west to east and north to south, and it is along those axes that the distilleries are located. They all place their worm tubs outside, for the reasons previously stated.

The bottom end of the worm condenser is brought once again into the still room and then connected to the Spirit Safe. The dimensions of the Spirit Safe are as follows: height = 1 *shaku*, 5 *sun*; length = 4 *shaku*; width 2 *shaku*. It is a brass box with panes of glass all around that can easily be removed when it is necessary to open the seal put in place by the Excise Officer.

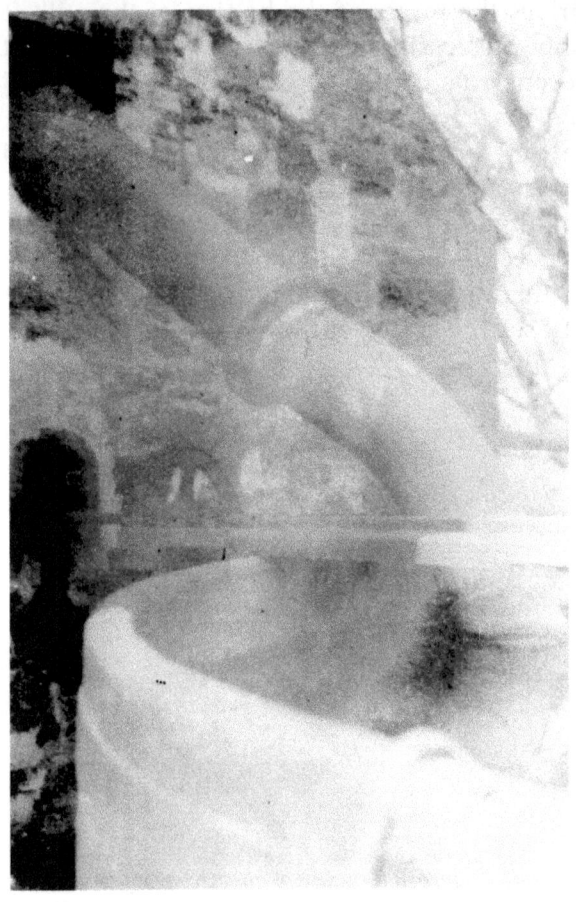

Photograph showing the way in which the pipe is led outside by knocking through the wall of the still room.

Worm tubs

The diagram on the next page shows a cross-section of the Spirit Safe. The box is arranged in three parts. There is the outlet for the low wines that come off the first distillation and then the outlets for the spirit, fore-shots and tailings obtained after second distillation.[20] Then, on the left, is the sample cylinder. The pipe from the wash still's worm is first connected to the whisky receiver (イ) and then, after passing through the metal mesh (ロ), is led to pipe (ハ). (ロ) is simply a metal copper mesh used to strain deposits produced in the process, such as sludge.

The pipe (ニ) is the outlet for the product of the second distillation.[21] (ト) is the pipe that passes to the whisky receiver and (チ) the pipe used for the fore-shots and tailings. After tests have determined that whisky of the required quality has been produced, the handle (ヘ) is turned and the spout (ホ) is turned in the direction of (ト), allowing collection of the whisky. In the case of the fore-shot and tails, it will be turned leftwards to (チ).

The sample cylinder (ル) is used for the low wines; when the meter displays 0 degrees, the distillation is stopped. (オ) is the cylinder used to test the whisky and, in the event that the liquor needs to be diluted down to 45 degrees, water is poured in via the vessel (カ), using the stopcock (ワ) to adjust the amount. The alcohol content is determined by means of the hydrometer's weights, while the quality of the product is determined by means of a test to see whether there is any clouding of the dilutant. (リ) is the cylinder cock — the handle is attached to the outside of the box, meaning that it can be engaged without opening the seal.

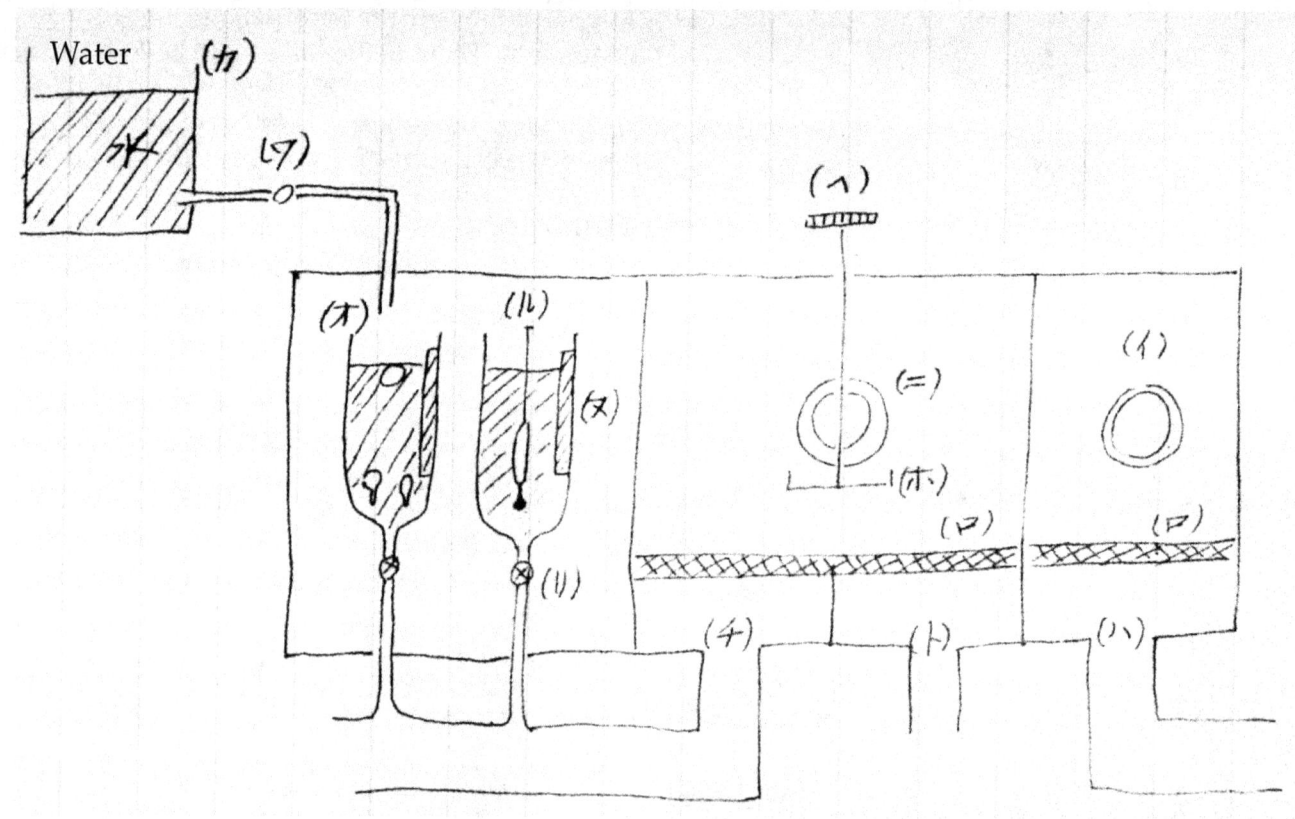

Cross-section of spirit safe.

1. The distillation process

There are two stills, each with a capacity of 100 *koku*, and they take a total of 540 *koku* of fermented wash from wash backs No.1-No.4. Firstly, 135 *koku* of wash from the first wash back is transferred to the wash charger and, after it has been inspected by the Excise Officer, two hours later it is poured into the still. The 135 *koku* are divided into two batches, 67 *koku* 5 *to* being the amount sent for first distillation. Into this 180 *monme* of soap is added in order to prevent bubbles forming. When coal is being added to the furnace, the mixer is turned simultaneously, thus mixing up the wash.

On average, the still can distil 11 *koku* of wash per hour, and therefore the first distillation can be completed in 6 hours. This includes the time needed to send the wash to the still and the time needed to remove the pot ale. Since this is the first distillation, all that we are aiming for is the collection of alcohol, so quality is not an issue. For that reason, when the meter shows that the alcohol in the Spirit Safe is at 0 degrees, distillation stops and the pot ale is expelled through the bottom of the tank (wash still).

The volume of low wines will be roughly ¼ of the total volume of the wash charge and will be of an average of 20 percent alcohol. This is transferred to a sealed receptacle,[22] and preparations made for the second distillation. Then wash still No. 1 continues distilling once again and if, during that time, enough has been obtained to send to the second still (the spirit still), that still is opened up and the second distillation gets underway. The capacity of still No. 2 is slightly less than that of still No. 1, but this is normal and at Hazelburn the two stills actually hold more or less the same amounts (100 *koku*). The construction of the stills is almost exactly the same, except that, since it handles not wash but alcohol that has already been distilled once, there is no worry that anything will stick to the base of the Spirit Still and burn, and so there is no need for mixing apparatus.

The volume of liquor supplied to this still for second distillation is about 1/3 of the total capacity, that is to say between 33 and 38 *koku*. At this stage, that which has been obtained is called the fore-shot and it contains 61-71 degrees of alcohol. It is not sent to a spirit receiver but to a separate feints receiver, where it will be mixed with the next batch. That which is obtained next is whisky and it is of 63-70 degrees. It is clear and has the distinct properties of 'new whisky'. When an equal quantity of water is added, a very slight clouding occurs – but this is unavoidable.

If a third distillation were carried out, obtaining a whisky with an alcohol content of 83 degrees, and then an equal quantity of water added to it, there will be no clouding. However, even whisky that has become cloudy will, after several years in storage, have improved in quality and, if drunk, it will be found that this cloudiness has added a special character to the taste of the whisky.

As for the question of determining quality, during distillation the lead technician will, on the permission of the Excise Officer, open the seal of the whisky receiver and continually taste the whisky, so one is not hampered by the constraint that one must wait until the distillation process is complete in order to be able to ascertain the quality of the whisky.

The simplest way of determining the quality of the whisky during distillation is firstly to take a sample of the fore-shot and add water to it until it reaches 45 degrees abv and then, if there is no cloudiness, the distillate may be re-directed to the Spirit Receiver.

This method also applies when changing the tails. If water is added until the liquid falls below 45 percent alcohol strength, cloudiness will occur. However, this is nothing to worry about. Of course, even when cloudiness does not appear in the dilution, an inferior spirit can still be distilled and so from time to time tests must be carried out in order to test quality of the aroma.

The test for cloudiness may be carried out without breaking the seal on the whisky receiver. The water pipe is driven through the top of the sealed box and the pipe's stopcock is affixed to the outside of the box, thereby enabling it to be freely adjusted. The leading edge of the pipe stretches exactly to the top of the sample cylinder, and both the hydrometer contained within the cylinder and a thermometer are used to determine the degree of dilution at 45 degrees.

That which is produced last of all in the distilling process is known as the 'tails'. This falls between 28 and 34 degrees. The technicians pay unceasingly close attention to see whether anything of inferior quality has been produced and, if so, will immediately send it to the tails receiver even if it does contain a sufficiently high alcohol content. The fore-shot and the tails are mixed and then discharged when the next distillation is carried out. It will be helpful to refer to the following chart.

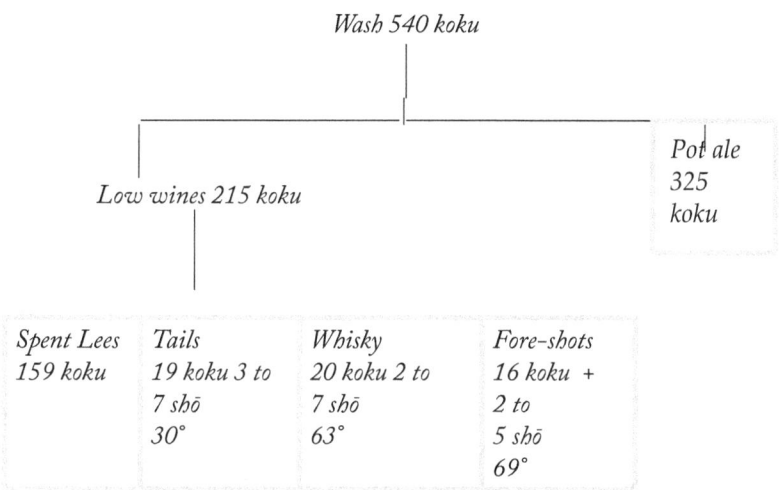

In much the same way that Japanese brewers use the specialist term *kosū* when making their calculations, when measuring spirit in the UK, a specialist term, namely "proof gallon", is used. The unit to which it refers is equivalent to 2 *shō* 5 *go* at 57°. Therefore, if we convert the products to proof gallons:

Fore-shots $\quad \dfrac{69\% \times 1625 \text{ }sh\bar{o}}{57 \times 2.5 \text{ }sh\bar{o}} = 787$

Whisky $\quad \dfrac{63 \times 2027}{57 \times 2.5} = 896$

Tailings $\quad \dfrac{20 \times 1937}{57 \times 2.5} = 272$

Total 1,955 proof gallons

That is to say, 1,955 gallons of spirit (48 *koku* 8 *to* 7 *shō* 5 *gō*) at 57° has been obtained. One may determine efficiency of output by comparing the number of proof gallons with the total amount of malted barley used.

2. Pot ale

The residue of the first distillation is known as "pot ale". It is constituted of the 60% of the wash sent to the still that remains in the bottom of the tank, and, if we take Hazelburn as our reference, were 67 *koku* of wash sent to the first distillation, 40 *koku* will remain as pot ale. For that reason, from the 540 *koku* of wash of the first distillation, a large quantity of pot ale — as much as 325 *koku* — will remain.

In the case of continuous distilling, the pot ale is sent to a sedimentation tank, where it is strained. Some of the clear liquid is then used in place of water for mashing, and the dreg, after it is dried, is sent to livestock farms. However, in the case of pot still distilling, the pot ale is not used, but is discharged just as it is via pipes as waste. Since Campbeltown lies on the coast, there is no problem associated with dealing with the pot ale in this way. Hazelburn lies at a distance of 8 *chō* from the sea. A concrete pipe with a diameter of 2 *shaku* has been laid between Hazelburn distillery and the loch, and this now takes away all the waste from the distillery. The clay mouth of the pipe is visible above ground at low tide and, despite the fact that the waste from all the distilleries flows into the sea, there are no adverse effects on the fish stocks. Indeed, as I mentioned in the Preface to this report, Campbeltown is one of the country's leading fishing ports.

For that reason, I cannot help drawing the conclusion that there must be some connection between the material discharged and the [health of] the fish. There is barely any acidity detectable to the tongue in the waste material, and so it must be recognised that the technicians employed at distilleries such as Hazelburn are in a much happier position than those in our company, who must grapple with this problem on a yearly basis. If the Japanese people could but develop refined tastes, and stop focusing their gaze on cheap whisky alone, the producers might pay more attention to their raw materials. That would then enable them to produce products with a special aroma and flavour that defy description, undetectable by means of chemical analysis. If that could be achieved, then, as a result of changes in the raw material used, fermentation times could be shortened and the acidity in the product made virtually undetectable. The waste discharged might then actually be something that the fishing and farming communities would welcome.

In Japan, companies such as ours have bought in burnt rice when fire has damaged storage facilities and we have had to collect wet barley from shipwrecks. Under such conditions, the amount of acidity on fermentation will be at the level of one or two percent, and the technicians will be under pressure to up the rate of production even if only in the smallest degree. They might even experience nightmares, in which they dream that they are under assault from bacteria. To compare our situation with that of the technicians at Hazelburn would be akin to comparing night and day.

That which remains as residue after the second distillation is known as "spent lees" and consists mostly of water and a small quantity of fusel oil, with nothing else in the way of solids. It is discarded together with the spent lees. About two-thirds of the liquid sent to the still remains at the bottom of the tank as spent lees, and so when a total volume of 215 *koku* of that liquor is distilled once more, 159 *koku* of spent lees is obtained. To sum up: 540 *koku* of wash is distilled and, on removing 325 *koku* of the first pot ale and 159 *koku* of the spent lees, 56 *koku* of spirit, fore-shots and tails remains.

This is a record of one half of the process as it is carried out at Hazelburn. In fact, double that amount is produced. That is to say, in a one-week period, 400 *koku* of malted barley is used to produce 1,080 *koku* of wash, and this is then distilled. Put together, the volume of new-make spirit, fore-shots and tails thus obtained is, at 57°, 97 *koku* 7 *to* and 5 *shō*. The fore-shots and the tails are included when calculating the amount of whisky produced. This is the production total for Week One and so, as no fore-shots or tails were available to be carried over, and amount of whisky produced is small. However, from the following week, the fore-shots and tails from Week One are added, and this naturally increases the volume of whisky produced. The fore-shots and tails are brought forward in this way from one week to the next.

When the fore-shots and tails are distilled for a second time, it is questionable as to whether an equal quantity of whisky is produced. However, in practice the volume does not alter much. For that reason, at Hazelburn, for example, 400 *koku* of malted barley will, in the course of a week, produce on average 97 *koku* of whisky at 57°.

Pot stills have no apparatus that enables the collection of fusel oil. One might think that, as the fore-shots and the tails are mixed into the next distillate, the amount of fusel oil contained therein would increase, but, in reality, an almost constant amount is carried over. When fusel oil is heated up repeatedly three to four times inside the still, whisky's characteristic flavour and aroma are created. Even when water has been added, it changes into a constituent element that produces hardly any cloudiness. Moreover, part of the fusel oil is discharged together with the spent lees.

3. The alcohol meter

Unlike in Japan, the term 'percent' is not used in the UK when referring to the degree of alcohol present in whisky. Instead, the term "proof" serves as the equivalent. Our 57° is the set figure that corresponds to "proof": alcohol exceeding this set figure is said to be "over proof", while that which falls below is said to be "under-proof".

A figure of 73.5 "over proof" equates to anhydrous alcohol, while a figure of 99.6 "under proof" represents water. "Proof" relates to the following: at a temperature of 51°F, the weight of spirit will be 12/13 of the weight of an equal amount of distilled water.

A foreigner such as myself struggles in the beginning to understand all this. For example, according to the restrictions imposed on its manufacture, whisky that is under 35% proof cannot be sold - to put it in terms that we Japanese would understand, this would equate to 36% volume. It is not just that the units applied are complicated, the meter used to test the whisky displays values that differ from "proof" values. The meter was invented by Sikes and is known as the Sikes' Hydrometer. It is made entirely of brass and beads are placed at the tip to do the measuring. The graduations of the scale start at the top with 0°, the very bottom is set as 10°. The weights range in units of 10-90, and there are nine of them. When at 15-16° the Sikes' Hydrometer displays 58.8 - this is 57% volume proof. Since it is necessary to convert to a temperature value, it must be admitted that this is an extremely complicated method of calculation. However, since the Sikes' Hydrometer is made entirely of brass, it has the advantage of being unbreakable. I

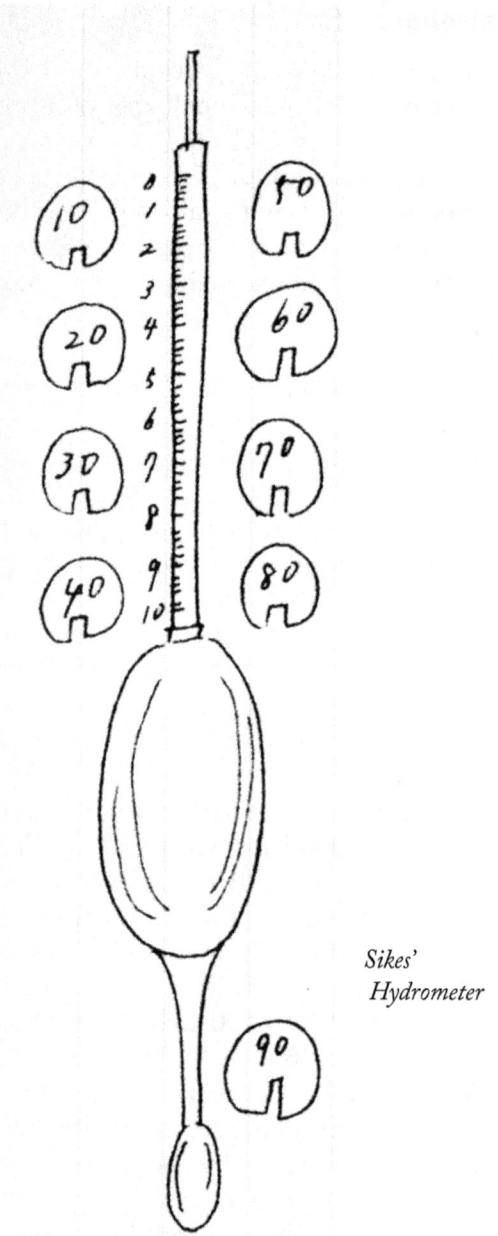

Sikes' Hydrometer

ought not to criticize the complexity of the alcohol meters used in the UK just because I personally find them inconvenient. It is simply that they don't use the same terminology we Japanese use in industry and commerce (i.e. *percent*), but rather rely on the number of degrees shown on Sikes' Hydrometer in order to determine the amount of alcohol present in the liquid.

For example, Westerners will criticise the Japanese for their use of chopsticks, while Japanese people will laugh scornfully at the use of knives and forks, which to us are troublesome to use. It all amounts to the same thing in the end and one must bear in mind that the important thing is whether the implement serves its purpose.

At the time of distillation, a glass Sikes' hydrometer and 3-4 glass beads, each in simple glass cylinders, are placed inside the Spirit Safe. Numbers are displayed, and so, if weight 11 is found to be floating, that will mean that the alcohol content is 11 over proof – this is what we in Japan would describe as 63%. If all the weights are floating, the degree of alcohol will be low and if all the beads have sunk then the liquor will contain a high degree of alcohol.

The sample cylinder and the floating weights. (A thermometer is shown on the left-hand side of the cylinder.)

Glossary

So far, I have not used a single Roman letter in this report.[23] It has been my intention to give translations directly into Japanese using Japanese script for the benefit of Japanese readers. However, there are a few short phrases that cannot be translated directly. Since I feel that this may hamper a proper understanding of the report, I hereby provide a list of the so-called "technical terms".

Malt barn	大麦発芽室
Steeping cistern	浸水池
Kiln	麦芽乾燥爐
Grist	麦芽を粉砕したるもの
Mash tun	糖化機
Rakes	糖化機内の攪拌用熊手
Husk	もみがら
Mashing liquid	前回に於ける最後の糖化洗滌液を次回の糖化用給水に代用するもの
Wort	糖化液
Draff	糖化残渣
Wort refrigerator	糖化液の冷却機
Wash	発酵醪
Wash back	発酵桶
Wash charger	醪溜
Still	蒸餾機
"　　*Pot still*	単式蒸餾機
"　　*Patent still*	連續式蒸餾機
Worm Condenser	冷却蛇管
Worm end	蛇管の端末

Safe	ウヰスキー蒸餾受け
Low Wine	第一回蒸餾液
Pot-ale	第一回蒸餾残渣
Wash still	第一回蒸餾機
Low wine still	再餾用蒸餾機
Spirit	蒸餾に當時のウヰスキーはこれを Spirit とも云い数年貯蔵した後 Whisky と云ふ
Faint [sic]	再餾の際 Spirit の前後に出る液と云ふ
" Fore shot	前餾一名初餾
" Tailing	餘餾一名尾餾
Spent-lees	再餾に於ける蒸餾残渣
Low wine receiver	第一回蒸餾液の容器
Spirit receiver	ウヰスキー容器
Faint [sic] receiver	再餾に於ける初餾及び餘餾の容器
Bate's saccharometer	Bate氏の発明せる検糖機
Sikes' hydrometer	Sikes氏の発明せるアルコールメーター
Proof strength	アルコール含有容量57%
Proof gallon	57%のもの二升五合
Over proof (O.P.)	57%以上のもの
Under proof	57%以下のもの

My personal view as to how we should implement distilling at Head Office in Japan

The ideal situation would be to install two stills, one for the first distilling and the other for the second. However, we might start off with one and use it for a double purpose. The structure should conform entirely to the Scotch whisky style, but, if the capacity of the still were to be reduced to 80 *koku*, then 35 *koku* of wash could be distilled at one time. As set out in the previous chapter on saccharisation, we could make 70 *koku* per mash and, by doing that twice, produce a total of 140 *koku*. That would mean separating the wash distillation process into four batches.

As to the question of whether the distilling should depend on direct or steam heat, I am of the opinion that this needs to be decided on the basis of the surrounding environment. Scotland is a nation that produces large amounts of coal and, so apart from the cost of extraction, no transportation costs are incurred. Relatively speaking, it is therefore rather inexpensive to produce. The distilleries have an ample supply of coal and almost all those engaged in pot still distilling use the direct heat method. That being said, there are a small number that do use steam.

From what I have seen, there are a large number of distilleries that do not have a boiler installed. The reason for this is that they employ the direct heat method for distilling and so all that they require is a means of producing hot water for the mash. To that end, they have 2-3 copper kettles — approx. 20 *koku* in capacity — and these are heated by coal, the water being then sent to the mash tuns through a pipe. Moreover, after the kettle has performed its purpose, it can be used to boil the water that will be used to clean the instruments and equipment. There is no need, therefore, to install a boiler – the lack of one does not have any adverse effect on the production process.

At Hazelburn, a 45 horsepower "Lancashire" boiler is installed. However, since only the direct heat method is used, the steam from the boiler is simply used to heat up the mashing liquid and the final batch of wort, the latter of which is heated so that it may serve as the next mashing liquid. This year for the first time steam has been required by a dryer for the purpose of drying the draff, and so the need for a boiler has arisen.[24]

From the point of view of our company, if distilling were to be conducted by the direct heat method and on a small scale, there would be no need to install a boiler. Since a boiler would be required if distilling were to be conducted by means of steam, it would be possible for steam to be supplied by the main plant to the distillery, provided the distillery were built directly adjacent to the main plant. However, as I recall, at present there is no surplus in the amount of steam produced at our main plant that might be diverted elsewhere.

If I were to state my personal opinion, I think that the nub of the matter lies in whether the distillery ought to be constructed adjacent to the main plant. It would undoubtedly be far and away more convenient to do so from the point of view of ease of management. However, after extensive consideration, I feel it must be recognised that, when it comes to the manufacture of whisky, the paramount consideration has to be the natural environment.

Turning to water, that blessing of nature — in any one-week period (6 days) Scotland's distilleries use an enormous amount. To illustrate this, we might take as an example a distillery producing 1,000 *koku* of spirit per week.

Water used[25]:
Wash	1,300 *koku*
Steeping barley	350 *koku*
Cooling the wort	1,500 *koku*
Cooling the distillate	5,000 *koku*
Cleaning	100 *koku*
Total	8,250 *koku*

Of the total, 1,500 *koku* of water for cooling the mash and 5,000 *koku* for cooling the wash is sent back to the tank upstairs and then recycled after being cooled by air. However low the estimate, one needs to be prepared for a consumption of at least 6,000 *koku* per week.

Despite the fact that the Campbeltown distilleries obtain water from a large loch situated in the hills of Argyll, shortages arise during the summer months, and production has to cease for a considerable length of time. Even in a country such as Scotland, which receives large amounts of rain, this state of affairs is still not unknown. Therefore, considering that the area around Sumiyoshi[26] is not favoured by natural resources, it would be illogical in the extreme to establish distilleries there, where water is scarcely obtainable, even if one digs wells.

If it were simply a matter of [normal] alcohol production there would be no problem, but to replicate true Scotch whisky the primary consideration must be the quality of the water and the second most important will be the quantity of that water.

It might not be possible to reconcile the ideal with the practical but, for my part, the ideal would be to conduct a thorough geographical survey and choose

1. A place with exceptional water quality and abundant water.
2. A place where barley may be easily accumulated.
3. A place where coal or firewood is easily obtainable.
4. A place with railroad links nearby.
5. A place not far from a freshwater stream.

If somewhere could be found that fulfils all the afore-mentioned criteria, I believe that establishing a branch factory in such a place could only add impetus to our future development. Since Japan has a lot of timber, even in places where there is a lack of coal, firewood can be used instead. I believe firewood would be the best alternative when it comes to the production of pot still whisky.

If there is a railroad nearby, then the whisky can, after it has been stored for several years, be taken out of storage and sent to the main plant in response to demand. Whilst having a branch distillery might at first glance appear uneconomic from the point of view of transport costs, when this is set alongside the advantages of having abundant water, a convenient means of wastewater disposal and an excellent final product, this pales into insignificance as a consideration. However, if, for reasons of cost, it be imperative that the distillery should be part of the main plant, then the still should be powered by steam, and that steam supplied from the main plant.

Steam produced from a boiler is far more economical than the direct heat method when it comes to the consumption of coal. In the case of the direct heat method, if 1,800 *kin* of coal is burned, only 2 *koku* 8 *to* 5 *shō* of whisky at 50% can be obtained. That is why at a distillery such as Hazelburn where nearly 100 *koku* of whisky are produced per week, the amount of coal used can be as much as 40 tons per week.

In large-scale operations, it can be profitable to install a boiler. In small distilleries, however, during the period from Monday to Wednesday, when distilling is not in progress, the boiler would need to be stoked in order to heat water for the mash. This is inconvenient and so the still is heated by the direct heat method and hot water is heated in an ordinary kettle.

As for distilling, if we assume that at our company in Japan, a total amount of 140 *koku* of wash were to be produced, the process would need to be split into four stages. The first distilling would amount to roughly one third of the total amount of wash, that is to say approx. 50 *koku* and, as it would then be distilled for a second time, the 50 *koku* would at that stage be divided and 25 *koku* distilled each time. In that way there would be no danger of bubbles flowing into the worm condenser. The distilled whisky would amount to 17 *koku* in volume at 50%. It would then immediately be transferred to barrels of 1 *koku* capacity and sent to storage. If the storage facilities are cramped, then it would be more efficient to produce whisky of 60% to be stored in barrels of 1 *koku* 5 *to*.

Both in Elgin and in Campbeltown, they produce 63% whisky for storage, while in the south of Scotland they store whisky of 71%. The question of whether the whisky going into storage is 71% or 50% will greatly influence the amount of storage area that is required.

The Storage of Whisky

Below, for reference, I set out one part of the Excise regulations relating to whisky in the UK.

Item: Whisky that is less than 45° and more than 71° may not be stored.

Item: Twenty-four hours prior to storage of whisky, the Excise office must be informed and the work must be carried out between the hours of 8 a.m. and 3 p.m.

Item: Within ten days of distilling having been completed, the whisky must all have been moved to store – however this does not apply in the case of continuous distilling.

Item: The use of barrels with a capacity larger than 3 *koku*, 7 *to* and 5 *shō* is not permitted.

Item: The export from storage of barrels containing less than 2 *to* 2 *shō* 5 *gō* is not permitted.

Item: Barrels being removed from storage must either be full or fall short of full capacity by less than 2 *shō* 5 *gō*. However, if intended for export overseas, the barrels may fall short of full capacity by as much as 5 *shō*.

Item: Whisky of more than five years that has been stored in first-filled barrels, together with whisky that has been blended with other types, can be taken out of storage to be used as a sample without being liable for Excise duty. However, all other whisky, even though it is to be used as a sample, is liable for Excise duty.

Item: Samples of whisky intended for export overseas are not liable for Excise duty.

Item: If there should be a fire at a warehouse, whisky that has been lost is not liable for Excise duty.

Item: If there should be a fire or some such other event in the warehouse, Excise duty already paid on whisky that has been lost will not be refunded.

Item: Whisky may be removed from the bonded warehouse under the following conditions:
1. For direct sale.
2. Transportation to storage facilities belonging to HM Customs and Excise prior to export overseas.
3. Transportation to another warehouse.

Scottish bonded warehouses are of two or three storeys but whisky is not stored at basement level. The reason for this is the damp nature of the ground. Above ground, the percentage of alcohol present in the whisky will be of a sufficient quantity to protect it against decay.

When one looks into the conditions in which alcohol is stored generally, one finds that it varies from country to country. For example, in France, it is essential that 'pure wine' - especially in the case of Bordeaux (alcohol content: 8-10 percent) - is stored below ground. The main reason for this is that the alcohol content is not high enough to protect the wine from becoming rancid through storage above ground during the summer months. Secondly, it is thought that the humidity of the conditions when wine is stored below ground imparts a special aroma and flavour to the wine. However, in Spain, white wine, as well as sherry and port, are all stored above ground.

This is because there is sufficient alcohol present at time of storage to preserve against their becoming rancid (19-23 percent). However, all the warehouses are single-storeyed buildings; there are none with two storeys. Whisky warehouses may be of two or three storeys, but it is considered that the most flavourful whisky is that which has been stored at ground level.

The barrels used to store the whisky are generally 1 *koku*, 5 *to* in capacity and the Exciseman will inspect both the number of barrels and their capacity prior to their going into the warehouse. Oak is used to make the barrels.[27] In the past, barrels were used for Scotch that had once been brought in from France or from sherry producers in Spain, but these are no longer sold and so a scarcity has arisen. For that reason, new barrels are now the norm and very few of the barrels now used are ones that once held sherry.

The Excise Duty on Scotch Whisky

Scotch whisky production has its origins in the 15th century and, from the beginning of 17th century, it gradually gained recognition. In the time of the famous Cromwell and his military government, an excise levy was placed on whisky. However, this was a mere 8 sen per 2 *shō* 5 *gō*. Thereafter, in the 18th century, duty was increased just at the time when whisky production was experiencing a sudden growth.

1856	2 *shō* 5 *gō*	4 yen
1861		5 yen
1885		6 yen
1912		7 yen 50 sen
1917		15 yen
1919		25 yen
1920		36 yen 25 sen

(The percentage of alcohol is taken as 57° and calculated at a rate of 1 shilling to 50 Japanese sen)

That is to say, today 1 *koku* of whisky at 57° is liable for a levy of 1,450 yen per *koku*, which constitutes an astonishing increase. If we consider this with reference to Japan, it is no wonder that whisky of 57° is liable for 60 yen in tax per *koku*, or even increased as far as 100 yen. Since such a precedent already exists in the UK, we must assume that the brewing and distilling industry in Japan will, in the near future, experience a similarly dramatic rise in the amount of tax levied.

Pre-war, the tax imposed in 1914 undoubtedly amounted to 200,000,000 yen and at that time whisky of 57° yielded 300 yen in Excise duty per *koku*. Ever since war broke out, the government has, in the main, curtailed the production of whisky in order that barley may be rationed. Owing to this interference in whisky production, the figure of 1,500,000 *koku* that stood in 1914 has gradually fallen. However, tax has increased so that today one *koku* gives 1,450 yen in tax revenue, as indicated in the previous table. This means that, while production has fallen, tax revenues have increased. A table of statistics relating to this year's whisky has not been published, so it is not yet possible to report any of the figures for the current year. However, the total amount of tax levied in any given year cannot be arrived at by merely multiplying the amount of tax due by the volume of whisky produced in that year. The reason for this is that whisky has to be stored for several years in the warehouse before it can be sold.

The amount of whisky stored in British warehouses at present awaiting imposition of Excise duty is an enormous sum, and can be said to reach as much as 5,000,000 *koku*. Last year, nation-wide Prohibition in the USA affected the situation in Britain, and last year some people called for a prohibition here, but this has yet to be implemented and would appear to be very far off. It is not just that to deny the Scots their whisky would be as unthinkable as denying the French their wine or the Spanish their sherry; not only would there be voices raised in vehement opposition from the general populace, the Government would be wary of such a plan in view of the effect it would have on the Treasury.

The Price of Whisky

It is extremely difficult to give an accurate report of the pricing of whisky. Roughly estimated, in 1914 the cost of whisky production was extremely low, with whisky of 57° costing in the region of 40 yen for one *koku* and, after three years in storage, the price before tax will have amounted to approximately 65 yen per *koku*. In 1918 this had jumped to 200 yen per *koku*. Today the price of barley has shot up and labour costs have also increased. As a result, the cost now stands at between 400 and 500 yen per *koku*.

If we take 400 yen to be the cost per *koku* and, calculating Excise duty to be 1,450 yen per *koku* for whisky of 57%, then the price per *koku* will amount to an astonishing 1,850 yen. These are all the wholesale costs, but in the case of retail sales or sales on the market (i.e. in bars), according to the price that was revised on 10th May, consumers will pay 1 yen 32 sen per gill or, in English money, two shillings and eightpence. A quarter gill (35 cc) costs 35 sen. Most drinkers drink a 35cc glass standing at the bar and to have to pay 35 sen comes as a huge shock to those who like to knock it back. Whisky sold by retailers has to be 36 percent according to government regulations. The actual cost of production generally speaking reflects a sharp rise in the cost of barley. Canadian barley sells for 35 yen per 1 *koku* 6 *to*, while Scots and Irish barley sell at 45 yen for the same amount.

Looking at the situation at Hazelburn Distillery, a week's supply of barley amounts to 280 *koku* (400 *koku* of malted barley)[28] and, if we assume a mixture of equal quantities of Scots and Canadian barley, then the average cost would be 40 yen per 1 *koku* 6 *to* and 7,000 yen in total for 280 *koku*.

If we add up the costs, including the cost of workers' wages and fuel for drying, up until the point at which the steeped barley has sprouted and then been dried, then we arrive at the following result: 1 *koku* 6 *to* of barley will cost 2 yen 50 sen, while the cost for the entire 280 *koku* will be 437 yen 50 sen.

The amount of coal used in a week will amount to a total of 40 tons. If we calculate a ton of coal at 25 yen, then the total cost will amount to 1,000 yen. The amount of yeast used per week is 104 *kan*, at a cost of 7000 yen.

It takes nine workers to carry out mashing and distilling and wages per worker per week amount to 50 yen, so that the total cost is 450 yen. If we add everything up, then the total production cost per week comes to 8,957 yen 50 sen.

Yen	
7,000.00	barley at 280 *koku*[29]
437.50	germination costs
1000.00	coal 40 tons
70.00	yeast
450.00	labour costs for 9 workers
8957.50	total

A by-product of whisky production is draff. This becomes fodder and sells for 50 sen per 2 *to*. Today, one week's worth of barley (280 *koku*) will give 450 *koku* of draff, giving a total income of 1,125 yen and if we subtract this from the previous total, we get a revised total expenditure for whisky production of 7,832 yen 50 sen.

Whisky production requires an average of 280 *koku* of barley to produce 98 *koku* of whisky at 57°. For that reason the actual cost per *koku* is 79 yen 93 sen.

7832.50 yen ÷ 98 = 79.93 yen

It is not that I investigated this by going into the distillery office (that would have been impossible). I listened to what a number of people had to say and, after ascertaining the price of all the commodities in question, I made the calculations myself. Therefore, there will probably be some errors, but I believe I have been able to present an accurate general overview. In addition to the figures given above, one also needs to add the salary costs of technicians, auxiliary workers and coopers.

Concerning cooperage at our own company

I think it goes without saying that the bigger the distilleries get, the greater the necessity to construct dedicated cooperages. I recall that Hokkaido produces good quality timber[30] and, if one looks into the geography of the country as a whole, I believe that, even without supplies from overseas, Japan has all the timber needed for cooperage. If, in the future, Scotch whisky production commences in Japan, it will be essential that we have enough barrels. To express this numerically, a new factory — even a small scale one - will produce 1,000 *koku* of whisky of 50° annually. Therefore, we will need 1,000 barrels with a capacity of 1 *koku* each year, and, for a three-year period, 3,000 such barrels will need to be made ready. In that case, assuming the whisky will be 60°, and that we will be using barrels with a capacity of between 1 *koku* 5 *to* and 2 *koku*, it follows that the number of barrels required will then be fewer. However, in the case of whisky that will be stored for at least three years (and some for as long as ten or fifteen years), more than 3,000 barrels will in fact be required. Since it will not be possible to obtain these barrels from overseas, new cooperages must be established so that we may look forward to the steady development of this enterprise.

Summary

When in the future I return to Head Office, I should like to put into practice what has partially been set out in my report on Hazelburn Distillery. However, to be on the safe side, I should like here to present a summary.

The mashing process requires that one mash comprise two extraction stages, and the amount of malted barley required for one mash is 30 *koku*, with 36 *koku* of water being added. Work begins at 7 a.m. on the first day and, after two and a half hours, the wort is passed through a cooling apparatus on its way to the fermentation room. Water is then added once again to the mash tun and the second mashing step[31] and cleaning are carried out, and after two hours that wort is sent to the fermentation room in the same manner as before, i.e. as was the case for the first batch of wort. There, the two batches of wort are then combined for fermentation. The cleaning liquor obtained by the third soaking is sent to a heated tank for use in the next round of mashing. As a result of the afore-mentioned process, the first mash concludes at 2 p.m. on the same day as it began. The second mash follows on and the rinsing liquid from the third batch of liquor is sent to the wash backs, and fermentation takes place.

The production process is therefore:

First extraction stage: 30 koku of barley giving 70 koku of wort – sent to 1st wash back

Second extraction stage: 30 koku of barley giving 70 koku of wort – sent to 2nd wash back

In this way, the first day's production finishes and, on the evening of the second day or on the morning of the third day, the product (fermented wash) of the first mash is split into two parts and 35 *koku* are distilled.

The time required for a single wash distillation may be estimated at 4 hours, and the four wash distillations that take place up to the point at which the second washback has been completed will require a total of 12 hours[32], and, at the very latest, will be completed by 7 p.m. on the evening of the third day. A total of 140 koku of wash produces 50 koku of low wines at 17 percent.

The low wines produced on the previous day are processed in two separate batches, commencing at 7 a.m. on the fourth day, and each time 25 *koku* are distilled for a second time. If one distillation is assumed to take 3 hours, then the second distillation will be completed by 1 p.m.

By means of this process, 8 *koku* 4 *to* of whisky at 50% is obtained, as well as 8 *koku* 6 *to* in the form of fore-shots and tails, with an average of 50%. The fore-shots and tails are, of course, calculated as whisky, meaning that the total obtained will be 17 *koku* at 50%. The reason for this is that, as I have said before, from the time of second distillation, the fore-shots and tails are carried over to the next distillation. For this reason, over the course of four days, from 60 *koku* of malted barley, 140 *koku* of wash is produced to give 17 *koku* of 50% whisky, which comes to 119 *koku* per month. Assuming the production period runs from the beginning of November until the end of March, the production total will be 599 *koku*.

However, this calculation allows for a great deal of leeway. If the workers work overtime and through the night and if, on top of that, the 2nd extraction stage is begun on the third day of the 1st extraction stage, the whole process may be shortened by one day. Therefore, I think it is safe to say that the annual production total will be no less than 1,000 *koku*.

For that reason, at Head Office we can expect to obtain 4,000 *koku* of new whisky annually, from a mixture of 3,000 *koku* of whisky distilled from sweet potatoes and 1,000 *koku* of pot still whisky.

As for the production period, in Japan the five month period between the beginning of November and the end of March is something we can adopt. I don't think we will be able to produce a superior product during the summer unless cooling apparatus is installed on a scale as prodigious as that employed by the *sake* industry.

In Scotland, for example, which has a cold climate, production can begin in October and continue for nine months until the following June. The three months of July, August and September correspond to Japan's climate in May but even so are rather unsuitable for whisky production and so those distilleries not engaged in continuous distilling shut during those months.

Principal reason – Climate
No 2 Insufficient water
No 3 Scarcity of barley
No 4 Livestock farms have enough grass not to require the supply of draff
No 5 Workers are engaged in farm work
No 6. The peat is dug in summer and laid out in the sun to dry

These are the reasons for the seasonal closure of the distilleries.

In Japan working during the summer months will have a deleterious effect on the germination of barley as well as upon fermentation and so, if possible, every effort should be made to produce the desired amount of whisky within a five-month period.

Appendix

The Question of Labour

As the world grows gradually more complex, whisky producers are no longer able to concentrate their labours solely upon research into barley and yeast. The question of labour has hitherto been disregarded but nowadays it is necessary to give it full consideration.

For that reason, here, at the end of this report on whisky, in addition to informing the company about questions of labour in the UK, I should also like to give my personal opinion on those issues for reference.

Since the end of the war, the first trade union to take strike action was that of the mineworkers and the next to do so were the railwaymen. In the end calls for strike action resounded throughout the land and even female shop workers have been swept up by the tide of unionisation. However, in none of the roughly 500 distilleries has there been a single day lost to strike action. When one considers the working conditions of the distillerymen, who work in a relatively clean environment, and whose work is carried out at a leisurely pace, it is clear how much their conditions differ from that of the miners, who must engage in heavy toil several feet underground without even a glimpse of sunlight.

It is because of this that miners' wages are the highest of the manual workers. They work a seven-hour day and a forty-hour week (half a day on Saturdays) and receive an average of 50 yen. With overtime, this rises to 70 yen per week. Distillerymen work a ten-hour day (half day on Saturday), making a fifty-five hour week, and receive a minimum of 30 yen and a maximum of 50 yen. Even the lowest paid among them can make 40 yen with overtime.

For that reason, from the point of view of the type of work done, compared with miners, distillerymen are treated in a much better way. Moreover, from the point of view of their character, distillerymen are more compliant and agreeable than the miners, and so have resisted the call to strike.

Wages in the distilleries have risen by 10 yen a week on last year. This took place as a result of the influence of the miners' strike. It would be true to say that the attitude of the miners is enough to change the fate of the whole country.

Miners' households can obtain coal for next to nothing, and so if any strike action they take to raise their wages leads to a steep rise in the price of coal, it will not bring about any increased hardship to the miners themselves. However, for other working class British families for the price of coal to rise in such a harshly cold climate where coal is needed both at the fireplace and in the kitchen, a steep rise in cost is as much of a shock as it would be for Japanese families if rice were suddenly to rise as steeply. Moreover, the manufacture of bread and other foodstuffs cannot, in the main, be carried out without coal. In the light of this, the enormous price rises occurring at present are extremely alarming. In the end this will have an effect on miners' families, too, leading to further calls on their part for wage rises, and it looks as if there will be no end in sight.

British trade unionism, which takes pride in being a pioneer of civilisation among the workers of the world, already draws upon a history of 100 years and has now raised wages as far as they can go.³³ When one considers that, even so, disputes unceasingly continue, it is no surprise that Japanese workers, who have been disregarded up to now, have begun to join in the global trend and are calling for a revolution of an unprecedented kind. However, it is regrettable that by thoughtlessly wielding the workers' only weapon - the strike – they take action without any definite ideology and simply follow along blindly.

Frankly speaking, there can be no doubt that Japanese workers are, as compared with those in the United States and Europe (in particular, the UK), both individually and in terms of their character, extremely inferior. Those who exercise control over the workforce should therefore take steps to improve the workers' character, and it will be imperative that they instil in their workers an awareness of the significance of their work. The fact that Japan was treated as a special case at the Washington Labour Conference has marred its standing, despite Japan's properly being regarded as one of the Five Great Powers. Those Japanese who took exception to this should go one step further and think over the fundamental question of why Japan had to be considered a special case in this way.³⁴

The Japanese remain a nation that has to work twice as hard as any other. The fact that today Japan has at last been recognised as the Land of the Rising Sun is solely the result of our having had to work harder than anyone else ever since the Meiji Restoration.³⁵ In particular, a country such as Japan which has an abundant population must set its sights on increasing production and bringing stability to the lives of its people.

My small suggestion would be to require workers to put in at least a ten-hour day.³⁶

Then they should be compensated in return for increased productivity with a share of the profits. As the ancients said, "A person with a fixed livelihood has peace of mind". No matter how tough their work, as long as their livelihood is guaranteed, they will consider themselves fortunate.

The working hours at pot still whisky distilleries are from 7 a.m. until 5 p.m. Within that ten-hour period, one hour is set aside for breakfast and lunch, making it a full nine-hour workday. The way in which they carry out their work reverently and calmly without even a set rest time seems in marked contrast to Japanese workers, who tend to be petulant and fret over trifles.

From the point of view of physique, their effectiveness at work is superior to that of our Japanese workers. One can say that their standard of living increases their effectiveness as labourers and so I feel that it is mistaken for Japanese capitalists to complain that for the workers to add raw fish to their dinner tables when they get a bit more money is an extravagance. In order to work hard, one must eat well. That is a biological principle.

In pot still distilleries, Saturday is taken as a half day and on Sunday no work is done at all. Wages are paid on a weekly basis. On Saturday afternoons and on Sundays when special attendance at work is required, overtime is paid at a percentage rate.

Now at Hazelburn Distillery, workers usually put in two hours overtime per day, working for a total of twelve hours, handing over to the night shift at 7 p.m. They are thus to be considered extremely hard-working.

At distilleries where continuous distilling is carried out, the process is more complex than is the case at pot still distilleries. Therefore, a greater number of workers is required, and they will work for eight hours per day in three shifts for each twenty-four hour period.

This is as set out below:

6 a.m. – 2 p.m.
2 p.m. – 10 p.m.
10 p.m. – 6 a.m. the following morning

They work on a rota system week to week so that in this kind of distillery there is no overtime work done.

Japanese workers are paid by the day so that if they take Sunday off, they will receive no pay for that day. This means that they are unable to take a day of rest with any peace of mind. If we look at companies in particular, public holidays are designated company closure days and so, since workers need to eat on holidays just the same as on work days but receive no pay when not at work, being given a day off must cause them little pleasure.

I think that two days off per month for the workers is quite sufficient. In my opinion they should have the two days and, as long as they are registered employees, should receive half pay on those days. In that way, the aim of providing days of rest will have been achieved.

Since Japan is not a Christian country, there is no need for Sunday to be a designated day of rest as it is in Europe and the United States. However, since the education system takes it as such, if workers were allowed to knock off an hour early on Sundays, those with school-age children would have the opportunity to enhance the happiness of their families and deepen the bond of affection between father and child. This would foster good feelings between the workers and the bosses.

As for wages, these are decided on the basis of the price of everyday goods and so, no matter how big the wage bill, if commodity prices shoot up, wages will have to do the same.

The wages of distillerymen at pot still distilleries may be estimated to be on average 170 yen per month, but this is a small sum when one considers that a loaf of bread costs 28 sen, an egg 16 sen, and 600 grams of ham 2 yen and, on top of that, one must pay 1 yen 50 sen for 10 *kan* of coal for the stove. These wages are therefore not generous enough to justify any envy on the part of Japanese workers, but to some extent it does clearly allow some money to be saved for old age and for any disasters or emergencies that may come along. For that reason, I believe that some means should be found of ensuring that wages allow Japanese workers to save for old age, not just pay for the daily necessities of life.

There are no regulations stipulating how much length of service should rewarded but at Hazelburn those who have worked for the company for five years or more receive the following as a bonus:

1 year   100 yen p.a.
6 years +   110 yen p.a.
7 years   120 yen p.a.

In this way, up to 23 years, the amount rises 10 yen per year.

23 years +	280 yen
24 years +	290 yen
25 years +	300 yen

This sum is calculated as of July 31st every year and so those who retire before that date in any given year will not receive the bonus for said year. If strike action has forced the closure of the distillery for eight days or more, bonuses for that year will all be cancelled, according to the rules.

As a result of this system, Hazelburn, which has a large number of experienced skilled workers who have been at the company for ever and a day.[37] They are so steeped in whisky from head to toe, they seem to spit flames. The company will immediately dismiss any worker who is not serious about the work without a second thought, but will hold onto hard-working men indefinitely. This is important from the point of view of guarding the secrets of production methods and ensures that the production process is speedily carried out.

I seem to recall that at our company in Japan there is no special reward for continued service but instead wages are raised annually, giving rise to a great disparity when long-serving workers are compared with newly-hired ones. In contrast, at the distilleries here in Scotland, there is very little difference between the rates of pay received by newcomers as compared to workers many more years' service. However, those who have given long service can rise to such positions as maltman, mashman, stillman, etc., and, on account of their qualifications, will receive a considerable rise in wages accordingly.

In short, we should not neglect labour problems, but rather investigate them constantly as a matter of course. If our Japanese workers were slightly better educated, all these problems could be peaceably resolved. For that reason, I take the view that, in companies employing large workforce, time should be spent on finding ways to enhance the workers' knowledge vis à vis their work and any other matters that require a common sense approach.

Recently in Japan it has been notable that, in the process of gobbling up Western ideas, the mistaken conclusion was reached that in Europe and the United States there is no consistency in the relationship bosses have with their various workforces. Ideas of equality and freedom are touted everywhere, but when it comes to their own workers, technicians and managers one does indeed find a comprehensive system of regulation in place.

This is one of the most important points that we Japanese have to study. I am of the opinion that we need to gain a proper understanding of the importance of obedience in exchange for freedom and equality, without recklessly conflating them.

The problem of pay and conditions

Company staff are neither capitalists nor workers but something in between the two. Therefore, they are caught in the middle. In Japan today, there is none so pitiable as the middle class. For that reason, I cannot help but assert the need for them to be properly compensated before embarking on a discussion of the treatment of manual workers. As to the question of how many such employees receive enough to ensure a settled livelihood, the number who do is extremely small. How can the majority support their wives and children? How can they ensure a secure old age? I think they must be undergoing extreme hardship. They cannot hope to enjoy the pleasures of life and will end up having lived a meaningless existence.

On arriving in the UK, I was immediately struck to realise that most of the population fully understands this point and they are intent upon increasing the happiness of each and every one. Not just in remote districts, but also in the towns, each week everyone stops work on the Sabbath, shuts the door and spends a whole day peacefully with the family. Everywhere on Sunday, the church bells ring out morning and evening, and men and women, young and old gather in church to hear the teachings of God and then they all take a walk as a family. This is indeed a way of life to be envied. I believe that affording the employee some leisure time and sufficient rest is an important element of the conditions of employment.

For manufacturers in general and in particular those who are technicians in brewing and distilling, it would not always be feasible for them to rest on Sundays as there will be the need to attend to fermentation or the propagation of yeast when production is underway. However, those involved in sales and other such office-based work do not need to be at the company when most other occupations have Sunday off.

In particular, the merchants of Osaka are rather slippery and even when orders can be made on days other than Sunday, they will call up on a Sunday. This is no proper way to conduct business. For that reason, if the staff have a day off on Sundays, they will make sure to take care of all orders on other days. Not only will this mean that profits remain unaffected, they will be able to conduct their business in a more regular fashion. However, if office workers take Sunday off, and, in addition, the two days per month designated as public holidays, this will mean that there are too many days taken each month. I believe that it will be necessary for the matter to be discussed and some kind of compromise arrived at.

If office workers can aim to work at maximum speed and efficiency, then they can feel free to return home, and those with families may enjoy a pleasurable evening together. This is what I hope for. This would not only mean that lives could be lived in a meaningful fashion, but would, in fact, be in conformity with the way that human beings were intended to live. Is that not so?

Sales methods

It goes without saying that sales strategy must be timely and quick, but one must also pay attention to the long term. There is no secret to making whisky; anyone can turn a hand to it. That is the reason that I am the first to have come to Scotland in order to do research as a prelude to commencing the production of authentic Scotch whisky in Japan. However, if, in the future, the industry is to flourish in Japan, many technicians will have to come to Europe. It will undoubtedly be the case that they will end up producing the same kind of whisky as we do after they return to Japan. It is not just whisky; it will not be especially difficult for them to produce every kind of alcoholic beverage and send it to market. We will naturally have to produce a superior product in order pre-empt our competitors, and will need to investigate other ways to sell our whisky.

Company housing for workers. This is the row of houses known as Millknowe Terrace.

My most fervent hope is that the company will sell directly to the market using the company trademark. If competitors emerge in the future, it will be likely that the Osaka wholesalers, those whom our company would feel to be the most skilful, will be the ones flocking in the direction of the new entrepreneurs. When that happens, it will not just be the case that if we suddenly attempt to sell directly to the market under our own brand we will have missed the boat, we will engender hostility among the wholesalers and end up falling between two stools.

However if, during this time when competitors are yet to emerge, we put this plan into practice, selling the product under a fixed brand, wholesalers will be forced to take supplies from us. As well as selling the brand to the public, we will also be selling to the wholesalers, and our business will increasingly flourish. Even if, in the future, the Osaka wholesalers all become enchanted with other brands, we will already be selling our brand to the public and so, not only will we remain unaffected by this turn of events, we will probably find that the wholesalers will have nothing to sell under their own brand and everything will be sold under ours.

Sale of Scotch whisky in the UK is as follows:

White Horse
Black and White

Also: Johnnie Walker

These are the well-known whiskies and, as well as being producers, they are also the merchants of their own brands. These same brands have enjoyed two hundred years of consumer confidence. These brands are, of course, of a superior kind, but the majority of distilleries are producing even better ones. However, their brands have not been considered superior by the public at large, and it seems that this has led to their being taken over by the companies mentioned above. For that reason, producing a superior product can only prove successful if accompanied by a good sales strategy.

It is easy to lose the goodwill of the Osaka wholesalers and difficult to gain the trust of the consumer. We must work hard to capture their interest and join forces with them. Since our company possesses great ambitions for achieving market domination in the East, we must improve the way we sell our products, rather than appeasing the wholesalers and sharing their fate. Moreover, we must, at the earliest opportunity, and in advance of our competitors, gain the public's trust in our products. In so doing, we will be laying down the foundations of everlasting stability.

The day of my return to Japan draws ever closer and, now that I must soon leave Scotland, in my dreams I have already rushed back to my homeland, where there is so much that I am longing to share with my dear colleagues at the company. Therefore, I now hastily conclude my report in the hope that there is nothing inaccurate to be found here.

I would like to express my gratitude in anticipation of your kind understanding.

Endnotes

1 Taketsuru uses the term "Elgin Range".
2 This is Beinn Ghuilean (pronounced Ben Gullion). The term "Argyll Range" would appear to be Taketsuru's own, adopted, perhaps, as a means of associating in his mind's eye the topography of Speyside with that of Kintyre.
3 i.e. denser barley
4 i.e. straw and husks
5 this is called the "chit", and the process is "chitting"
6 known as "acrospires"
7 Nada Gogō is the collective name given to five famous *sake*-producing villages, clustered along the coastal fringe of Hyogo Prefecture between Kobe and Nishinomiya. This area produces a significant proportion of all the *sake* produced in Japan.
8 i.e. not natural drying by the sun
9 Taketsuru here uses the foreign loan word *puropērā*.
10 i.e. from the previous mash
11 *mirin*: a sweet *sake* used in Japanese cuisine
12 The names given to the various species of oak and pine vary according to region. It is difficult to determine exactly to which species Taketsuru is referring, but at the time Taketsuru was writing, Oregon pine was the principal material used in the manufacture of wash backs.
13 This is called "bubbing the yeast".
14 A traditional type of Japanese pickle, named after the prefecture in which it originated.
15 Professor Tsuboi taught at what was then Osaka Technical High School, where Taketsuru was a student prior to his joining Settsu Shuzō. Prof. Tsuboi was an authority on zymology and Taketsuru recalled in his memoirs that Tsuboi was a charismatic figure.
16 Here a table showing the washback filling programme was inserted into the original notebook. It has been added to this translation as an appendix on page 72. N.B. some of the values do not appear to reconcile.
17 See Note 12
18 In fact some distilleries conduct both types of distillation production process on the same premises.
19 The original text would appear to be misleading here and so a portion of it has been omitted.
20 Taketsuru uses the word "whisky" here, but, as he notes in his own Glossary, at this stage the appropriate word is "spirit", as "whisky" is not used until the spirit has been stored for a number of years.
21 the feints
22 namely the low wines receiver
23 Original says "English letter", but "Roman letter" is more accurate.
24 Elsewhere in the text, Taketsuru mentions that the Campbeltown distilleries all send their draff to a central drying facility. However, by 1920 Hazelburn had its own draff drying facility on site.
25 It appears that Taketsuru might have underestimated his water requirements.
26 Sumiyoshi is the name of the area of Osaka in which Settsu Shuzō, the company that had sent Taketsuru to Scotland, was based.
27 Owing to regional variations in terminology, there is some room for debate as to exactly which species of tree Taketsuru refers.
28 These figures may have been erroneously transposed.
29 See Note 28
30 Owing to regional variations in terminology, there is some room for debate as to exactly which species of tree Taketsuru refers. Possible candidates are oak and pine, so the term "timber" has been used here as it covers both.
31 Traditionally called the "second water".
32 This figure should be 16.
33 Taketsuru gives the impression of his having been a liberal-minded man, keen to promote the well-being of ordinary workers. If his view on wages here appears pessimistic, it should be understood that he was writing in the context of the times. It would not be at all evident at that time that workers might secure even greater improvements in pay and conditions.
34 The term Five Great Powers refers to the five main signatories to the Treaty of Versailles: the United States, Great Britain, France, Italy and Japan, all allies

during World War I. In November 1919, the inaugural conference of the International Labour Organization (ILO) was held in Washington. The ILO had been formed in order to define and standardize workers' rights internationally. While Western nations pressed for standards to be raised, countries whose economies were weaker, such as China and India, were allowed to maintain lower standards. Japan had to decide to which group it would seek to belong. See "Who Speaks for Workers? Japan and the 1919 ILO Debates Over Rights and Global Labor Standards." Dorothy Sue Cobble. *International Labor and Working Class History.* No. 87, *Spring 2015 pp. 213-234.* The proceedings of the ILO Conference and its repercussions were widely reported in the British press both in 1919 and 1920. Again, Taketsuru's remarks should be understood according to their historical context. Half a century after the Meiji Restoration, Japan still considered its position in the world to be extremely precarious. Workers' education and the reform of working practices were held to be key to the development of industry and, by extension, the creation of wealth and consolidation of the modern nation state.

35 In 1868, the office of shogun was officially abolished and nominal power restored to the Emperor.

36 One of the main topics debated at the Washington Labour Conference was the length of the working day. Japan's Labour Delegate, Matsumoto Uhei, pressed for it to be limited to eight hours, but was unsuccessful. Taketsuru would have been aware of that debate. Despite his apparent lack of support for the introduction of the eight-hour day, from his remarks elsewhere in the notebooks, it would appear that Taketsuru's sympathies lay with the trade unionists, who asserted, "Labour is not a commodity…Workers are… not to be bought and sold." ("Who Speaks for Workers? Japan and the 1919 ILO Debates Over Rights and Global Labor Standards." Dorothy Sue Cobble. *International Labor and Working Class History. No. 87, Spring 2015 p. 218.*)

37 The implied meaning here is that they are loyal, but also stubborn and set in their ways.

Further Reading

Barnard, Alfred. *The Whisky Distilleries of the United Kingdom*. 2008 ed. Edinburgh: Birlinn Ltd, 2008.

Checkland, Olive. *Japanese Whisky, Scotch Blend: Masataka Taketsuru, the Japanese Whisky King and Rita, His Scotch Wife*. Dalkeith: Scottish Cultural Press, 1998.

Glen, Iseabal Ann. 'An Economic History of the Distilling Industry in Scotland 1750-1914'. Ph. D. Thesis, University of Strathclyde, 1969.

MacDonald, Susan L. 'Trade and Economic Development in Eighteenth Century Campbeltown', Ph.D. Thesis, University of Edinburgh, 1982.

Martin, Angus. *Campbeltown Whisky: An Encyclopaedia.* : The Grimsay Press, 2020.

Nettleton, J.A.. *The Manufacture of Whisky and Plain Spirit*. Aberdeen: G. Cornwall and Sons, 1913.

Stirk, David. *The Distilleries of Campbeltown: The Rise and Fall of the Whisky Capital of the World*. Glasgow: Angels' Share, 2005.

The Campbeltown Book. Kintyre Civic Society. 2003.

Wash Back Filling Programme

Ball = Balling Scale
SG = Specific Gravity

Item / Time	Wash back 1				Wash back 2				Wash back 3				Wash back 4			
	Bate's Saccharometer	SG	Ball	Temp	Bate' Saccharometer	SG	Ball	Temp	Bate's Saccharometer	SG	Ball	Temp	Bate's Saccharometer	SG	Ball	Temp
Mar 1 Mon 10.00	46°	1.128	29.5°	22°C												
18.00	35°	1.098	23°	28°C	45°	1.125	29°	22°C								
Mar 2 Tue 02.00	10°	1.028	7.2°	32°C	35°	1.098	23°	28°C	41°	1.114	27°	23°C				
10.00	3°	1.008	2.0°	33°C	12°	1.035	8.5°	32°C	23°	1.065	16°	30°C	35°	1.098	23°	23°
Mar 3 Wed 10.00	0			32°C	2°	1.007	1.8°	33°C	5°	1.015	3.7°	34°C	10°	1.028	7.2°	32°C
22.00					17.00hrs 0			32°C	0			33°C	2°	1.007	1.8°	33°C
Mar 4 Thur 04.00													0			33°C

Japanese –UK – US – OIML Measurement Conversion Tables

Following are tables of length, area, weight, volume, density and spirit strength, temperature and currency which readers may convert the unit measurements stated by Masataka Taketsuru throughout his text into ones with which they are more familiar.

Weight and volume were closely linked in pre load cell days with a particular volume often being the basis, together with an assumed or measured density, to determine weight. Taketsuru records that a bushel is 2 "to"1"gō" and an (Imperial) gallon is 2 "shō "5"gō" 2 "shaku".

Taketsuru made detailed calculations of the output of Hazelburn Distillery using the factor 57% (actually 57.15) to convert the UK proof system to ABV (Alcohol by volume).

The tables are approximated arbitrarily in order to be useful, rather than to any consistent rule, such as two decimal places.

Alan G. Wolstenholme

Length	Japanese		Metric		UK Imperial		USA
	1 bu		3 mm		0.12 in		0.12 in
	1 sun		3 cm		1.2 in		1.2 in
	1 shaku		30 cm		1 foot		1 foot
	1 ken		1.8 metres		6 feet		6 feet
	1 chō		109 metres		120 yards		358 feet
Area	1 tsubo		3.3. sq. metres		4 sq. yards		35.6 sq.feet

Weight	Japanese	Metric	UK Imperial	USA
	1 monme	3.75 grams	0.13 ounces	0.13 ounces
	1 kin	600 grams	1.24 lbs	1.24 lbs
	1 kan	4 kgs	8.27 lbs	8.27 lbs

Volume	Japanese	Metric	UK Imperial	USA
	1 shaku	1.8 cl	0..63 fluid ounce	0.61 fluid ounce
	1 gō	18 cl	0.32 pints	0.38 US pints
	1 shō	1.8 litres	3.17 pints	1.9 US quarts
	1 to	18 litres	3.97 gallons	4.77 US gallons
	1 koku	180 litres	39.7 gallons	47.7 US gallons

Density and Spirit Strength	ABV	10	20	30	40	50	60	70	80	90	100
UK (Sikes)	0 (97UP)	17.5	35	53	70	88	106	123	140	159	175 (75OP)
US Proof	0	20	40	60	80	100	120	140	160	180	200

Temperature											
Centigrade	0	10	20	30	40	50	60	70	80	90	100
Fahrenheit	32	50	68	86	104	122	140	158	176	194	212

Currency				
	Japanese	Sterling (£sd)	Sterling (decimal)	US$ (1920)
	1 yen (100 sen)	2 shillings 2 d [pennies]	£0. 11 [pence]	$0.44 cents

www.ingramcontent.com/pod-product-compliance
Lightning Source LLC
Chambersburg PA
CBHW081205170426

43197CB00018B/2933